T0257810

Encyclopedia of Alternative and Renewable Energy: Solar Energy Applications

Volume 20

Encyclopedia of Alternative and Renewable Energy: Solar Energy Applications Volume 20

Edited by **Catherine Waltz and David McCartney**

New York

Published by Callisto Reference,
106 Park Avenue, Suite 200,
New York, NY 10016, USA
www.callistoreference.com

**Encyclopedia of Alternative and Renewable Energy:
Solar Energy Applications
Volume 20**
Edited by Catherine Waltz and David McCartney

International Standard Book Number: 978-1-63239-194-0 (Hardback)

Contents

Preface

This book on solar energy applications presents a broad overview on various aspects and applications of this renewable source of energy. It provides evidences of high thermal efficiency of the gravitational draught using concentrated solar heating, new technologies of generating organic fuels using solar heating, new sorts of photovoltaic cells, long term application of thermal solar power plants, the efficiency of thermal storage and solar power utilization in Niger. The readers will find the accompanying illustrations and pictures as a pleasing experience which will ease the text assimilation and make it an attractive source of information.

Various studies have approached the subject by analyzing it with a single perspective, but the present book provides diverse methodologies and techniques to address this field. This book contains theories and applications needed for understanding the subject from different perspectives. The aim is to keep the readers informed about the progresses in the field; therefore, the contributions were carefully examined to compile novel researches by specialists from across the globe.

Indeed, the job of the editor is the most crucial and challenging in compiling all chapters into a single book. In the end, I would extend my sincere thanks to the chapter authors for their profound work. I am also thankful for the support provided by my family and colleagues during the compilation of this book.

Editor

Fuel Production Using Concentrated Solar Energy

Onur Taylan and Halil Berberoglu

Additional information is available at the end of the chapter

1. Introduction

Limited reserves of fossil fuels and their negative environmental effects impose significant problems in our energy security and sustainability. Consequently, researchers are looking for renewable energy sources, for instance solar energy, to meet the energy demands of a growing world population. However, terrestrial solar energy is a dilute resource per footprint area and is intermittent showing substantial variability depending on the season, time of the day, and location.

One strategy to overcome these drawbacks of solar energy is to concentrate and use it for cleaning and upgrading dirty fuels such as coal and other hydrocarbons or converting renewable feedstocks such as biomass into carbon-neutral solar fuels. In this way, the intermittent and dilute solar energy can be concentrated and stored as a chemical fuel which can be easily integrated to our existing energy infrastructure. These advantages of solar fuels produced with concentrated solar radiation make them an attractive solution in our quest for renewable and clean fuels. Figure 1 shows the energy potential and carbon emissions by most commonly used fuels along with solar hydrogen.

Most common and available methods for solar fuel production are thermolysis, cracking, reforming, gasification and through thermochemical cycles. All these methods require high temperatures to produce solar fuel. Therefore, in these methods, there are some qualities of the feedstock or the reactor that need to be satisfied to attain high temperatures and efficient solar fuel production. For instance, the physical size and porosity of the feedstock play an important role. As the surface area-to-volume ratio of the feedstock increases, more reaction sites will be available for the reaction to occur, which increases the process efficiency. The feedstock should also have a narrow bad gap to lower the energy requirement for chemical process. Additionally, the material on the reactor walls should have high optical absorption to increase the temperature of the reactor and withstand high temperatures, and the window material should have high transmissivity to let the solar energy in to the reactor. More detailed property requirements are given by Nowotny *et al.* [1].

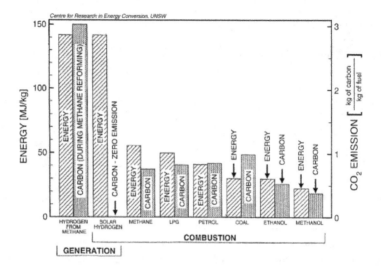

Figure 1. Comparison of different fuels in terms of their energy produced and CO_2 emission [1].

This review chapter consists of four sections. Following the introduction, the second section "Concentrated Solar Fuel Production Methods" reviews the different routes of producing solar fuels according to the feedstock material used in the processes. These include (i) thermolysis of water, (ii) thermochemical cycles, (iii) cracking of gaseous hydrocarbons, and (iv) gasification and reforming of coal and biomass. These methods are compared with each other based on their temperature, pressure, thermodynamic efficiencies, and by-products. The third section "Concentrated Solar Reactors" provides a comprehensive review of different concentrated solar reactor designs reported in the literature. This section first reviews the current solar concentration methods and describes in detail the effects of concentrating factors on the heat flux and temperatures that can be achieved. Then, the section describes the design and basic principles of operation of different solar reactors, their applicability for the different methods described in the preceding section, and their temperature and pressure capabilities. Moreover, the section summarizes the reported solar to fuel conversion efficiencies of each design. Finally, the chapter ends with the conclusions and outlook of fuel production with concentrated solar energy outlining the challenges, new research directions and novel applications.

2. Concentrated solar fuel production methods

This section describes different methods of producing solar fuels according to the feedstock material used in the respective processes.

2.1. Thermolysis of water

The term "thermolysis of water" refers to the thermal decomposition of water molecules into hydrogen and oxygen gases. Historically, due to high availability and simple molecular form of water, researches on solar fuel production started with direct hydrogen production by thermolysis of water using solar energy as,

$$H_2O \rightarrow H_2 + \frac{1}{2}O_2$$
$$\Delta H_{300K} = 286 \; kJ/mol \tag{1}$$

The reaction given in Equation (1) is an endothermic process, i.e., it requires energy to break the bonds. However, breaking all the bonds in water molecules requires temperatures as high as 2500 K [2]. At lower temperatures, partial decomposition occurs. Although it is possible to reach 2500 K with concentrated solar energy, the reactor where this process takes place shows material issues related to high temperatures. Additionally, after the dissociation of water molecules, hydrogen and oxygen gases require separation at high temperatures in order to prevent back-bonding, i.e., reproduction of water molecules with an exothermic process. Some solutions include cooling the reactor down by injecting a gas or expanding these gases through nozzle at the end of the reactor [2, 3]. Other solutions include using double or tubular membranes or using multi-stage steam ejectors to lower the exit pressure [4]. However, these solutions further reduce the efficiency of the process, and thus no commercial plant using this technology exists.

2.2. Thermochemical cycles

Some metal oxides are reduced using solar energy since metals provide good storage and transport of energy, such as solar energy. Such metal oxides include, but not limited to ZnO, MgO, SnO_2, CaO, Al_2O_3 and Ce_2O_3. The reduction step of these metal oxides is generally followed by an oxidation step at lower temperatures than reduction step in order to produces solar fuel, mainly hydrogen. The reduced metal oxides generally react with CO_2 or steam. If steam is used in oxidation that step is called hydrolysis. The thermochemical cycles of different metal oxides are generally compared based on their temperature requirements for the reduction step, the reaction or dissociation rates and reaction kinetics.

ZnO is one of the most popular oxides mainly due to its abundance and relatively low temperature requirement for complete dissociation when compared to other metal oxides. Additionally, since ZnO is a simple metal oxide, it does not undergo multiple reactions before its full dissociation. The dissociation of ZnO occurs as according to,

$$ZnO \rightarrow Zn + \frac{1}{2}O_2$$
$$\Delta H_{2000K} = 546 \; {}^{kJ}\!/\!_{mol} \tag{2}$$

The complete dissociation of ZnO to Zn requires temperatures higher than about 2300 K whereas, for instance, the dissociation of MgO as another simple metal oxide requires about 3700 K at atmospheric pressures [3, 5]. As in water thermolysis, partial dissociations can occur at lower temperatures. Although hydrolysis of zinc is exothermic as given by Equation (3), only 24% of Zn could be oxidized to produce H_2 at a reactor temperature of 800 K and an atmospheric pressure [6].

$$Zn + H_2O \rightarrow ZnO + H_2$$
$$\Delta H_{300K} = -62 \; {}^{kJ}\!/\!_{mol} \tag{3}$$

Figure 2 shows the overall process of hydrogen production from zinc-oxide.

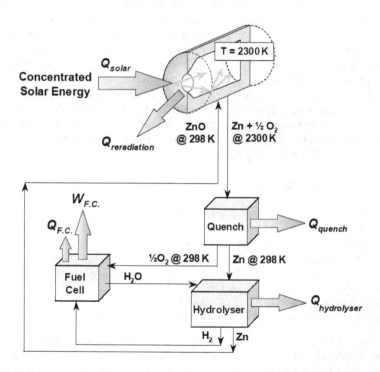

Figure 2. Flowchart for thermochemical hydrogen production from zinc-oxide using concentrated solar energy [5].

As an alternative to ZnO reduction, Abanades *et al.* [7] proposed SnO_2 reduction using solar energy. Once the SnO_2 is reduced to SnO in gaseous form using solar energy at temperatures nearly 1600°C, hydrolysis of SnO with steam at about 550°C and ambient pressure takes place in another step to form hydrogen gas as,

$$SnO_{2(s)} \rightarrow SnO_{(g)} + \tfrac{1}{2}O_2$$
$$\Delta H_{1873K} = 557 \; kJ/mol \tag{4}$$

$$SnO_{(s)} + H_2O_{(g)} \rightarrow SnO_{2(s)} + H_2$$
$$\Delta H_{773K} = -49 \; kJ/mol \tag{5}$$

The advantages of SnO_2/SnO reduction when compared to ZnO/Zn reduction are that (i) the SnO_2-to-SnO conversion can be increased in Equation (4) by decreasing the pressure of the solar reactor which increases the overall conversion efficiency [7] (ii) SnO has higher melting and boiling points when compared to those of Zn, so that quenching rate of SnO is not as important as of Zn [7] (iii) in ZnO/Zn dissociation, Zn needs to be quenched rapidly below its condensation temperature to prevent recombination, while this is not the case with SnO_2/SnO system.

There are some other metals that can be reduced with faster reaction kinetics such as Ce_2O_3. However, the reduction of Ce_2O_3 to CeO_2 starts at temperatures higher than 2300 K [8, 9]. Full dissociation requires higher temperatures. This requirement of high temperatures creates some material limitations on the material of the reactor and increases the cost of the reactor significantly. Although there are some lab-scale prototypes of Ce_2O_3/CeO_2 solar reactor, it is not preferred due to these limitations and high cost.

Another research was also started with producing hydrogen gas from hydrogen sulfide, H_2S, as,

$$H_2S \rightarrow H_2 + \tfrac{1}{2}S_2$$
$$\Delta H_{300K} = 91.6 \; kJ/mol \tag{6}$$

Hydrogen sulfide is a toxic by-product gas of sulfur removing process from natural gas, petroleum and coal. Thermal decomposition of hydrogen sulfide requires about 1800 K [10]. It is advantageous over the other metal oxide reduction processes discussed above since this thermochemical process is only a one-step process that does not require additional oxidation step to produce hydrogen. Additionally, the temperature requirement for dissociation is lower than that for the direct water thermolysis. However, the product gases need to be cooled down after the dissociation as in the water thermolysis or other metal oxide reduction processes [11]. Some studies showed that the temperature of reduction could be reduced to about 1500 K, and they showed that the reproduction of hydrogen sulfide is unimportant below 1500 K [3, 12, 13].

In general, the solar chemical process is a clean way to produce hydrogen without any carbon prints. Therefore, the hydrogen as a product of the solar chemical process can be used in fuel cells directly as it is pure. The solar chemical reduction step of the process produce nanoparticles with high surface area to volume ratio, e.g., Zn, SnO which also create additional reaction centers for the hydrolysis to occur [7]. Therefore, the oxidation or hydrolysis occurs fast due to high mass transport of gases in the solid phase [7]. As in the other dissociation processes, the products of the dissociation also need to be cooled in order to prevent re-oxidation. Sandia National Laboratories of US released a comprehensive report on the thermochemical cycle selection with initial selection for further research [14], and Table 1 summarizes the studied thermochemical cycles [15].

2.3. Cracking of gaseous hydrocarbons

The term "solar thermal cracking" or "solar cracking" is used for thermal decarbonization of natural gas or other hydrocarbons. As a result of cracking, hydrogen, carbon and other possible products are formed without CO_2 emissions. Therefore, this process is another method for clean fuel production. Cracking requires high temperatures of about 1500 K [16] that can be reached using concentrating solar collectors. For example, Maag et al. [17] tested a concentrated solar collector with a concentrating factor of 1720, and obtained a maximum temperature of 1600 K within the solar cavity reactor. In general, the advantages of solar cracking are the increase in value of feedstock using solar energy, pure and uncontaminated products and no CO_2 emission [16].

As being the simplest hydrocarbon and the main constituent of natural gas as given in Table 2, methane has been mainly considered for solar cracking. Chemical reaction of evolution of carbon black and methane is given in equation (7) [18, 19]. The kinetic mechanism of methane cracking at 1500 K and atmospheric pressure was proposed as [20, 21],

$$2CH_4 \rightarrow C_2H_6 + H_2 \rightarrow C_2H_4 + 2H_2 \rightarrow C_2H_2 + 3H_2 \rightarrow 2C_{(solid)} + 4H_2 \qquad (7)$$

Another important aspect of producing hydrogen and carbon black (solid carbon) is their market values. Hydrogen and carbon black have a market value of about \$135 billion per year and between \$7 and \$11 billion per year depending on the grade of the carbon black in the world, respectively [22].

Under an EU project named SOLHYCARB, a 50-kW$_{th}$ indirectly heated, cavity type solar reactor was developed for methane cracking [8]. Its 10-kW$_{th}$ prototype was built and tested using natural gas, and 97% conversion was obtained with a maximum temperature above 2000 K under concentrated solar irradiation of 4 MW/m^2 [23]. The difficulties that prevent this technology to become commercial are mainly the cost of the reactor and the complicated flow pattern inside the reactors. For example, in order to prevent particle accumulation on the window, some inert gas is introduced to the reactor with high flow rates and pressures, or indirectly heated solar reactors are used which decreases the solar-to-fuel conversion efficiency and further increase the cost.

Cycle	Reaction Steps
High Temperature Cycles	
Zn/ZnO	$Fe_3O_4 \xrightarrow{2000\text{--}2300^\circ C} 3FeO + \frac{1}{2}O_2$ $3FeO + H_2O \xrightarrow{400^\circ C} Fe_3O_4 + H_2$
FeO/Fe$_3$O$_4$	$CdO \xrightarrow{1450\text{--}1500^\circ C} Cd + \frac{1}{2}O_2$ $Cd + H_2O + CO_2 \xrightarrow{350^\circ C} CdCO_3 + H_2$ $CdCO_3 \xrightarrow{500^\circ C} CO_2 + CdO$
Cadmium carbonate	$CdO \xrightarrow{1450\text{--}1500^\circ C} Cd + \frac{1}{2}O_2$ $Cd + 2H_2O \xrightarrow{25^\circ C, electrochemical} Cd(OH)_2 + H_2$ $Cd(OH)_2 \xrightarrow{375^\circ C} CdO + H_2O$
Hybrid cadmium	$Mn_2O_3 \xrightarrow{1400\text{--}1600^\circ C} 2MnO + \frac{1}{2}O_2$ $2MnO + 2NaOH \xrightarrow{627^\circ C} 2NaMnO_2 + H_2$ $2NaMnO_2 + H_2O \xrightarrow{25^\circ C} Mn_2O_3 + 2NaOH$
Sodium manganese	$Fe_{3-x}M_xO_4 \xrightarrow{1200\text{--}1400^\circ C} Fe_{3-x}M_xO_{4-y} + \frac{y}{2}O_2$ $Fe_{3-x}M_xO_{4-y} + yH_2O \xrightarrow{1000\text{--}1200^\circ C} Fe_{3-x}M_xO_4 + yH_2$
M-Ferrite(M = Co, Ni, Zn)	$H_2SO_4 \xrightarrow{850^\circ C} SO_2 + H_2O + \frac{1}{2}O_2$ $I_2 + SO_2 + 2H_2O \xrightarrow{100^\circ C} 2HI + H_2SO_4$ $2HI \xrightarrow{300^\circ C} I_2 + H_2$
Low Temperature Cycles	
Sulfur-Iodine	$H_2SO_4 \xrightarrow{850^\circ C} SO_2 + H_2O + \frac{1}{2}O_2$ $SO_2 + 2H_2O \xrightarrow{77^\circ C, electrochemical} H_2SO_4 + H_2$
Hybrid sulfur	$Cu_2OCl_2 \xrightarrow{550^\circ C} 2CuCl + \frac{1}{2}O_2$ $2Cu + 2HCl \xrightarrow{425^\circ C} H_2 + 2CuCl$ $4CuCl \xrightarrow{25^\circ C, electrochemical} 2Cu + 2CuCl_2$ $2CuCl_2 + H_2O \xrightarrow{325^\circ C} Cu_2OCl_2 + 2HCl$
Hybrid copper chloride	$2CH_4 \rightarrow C_2H_6 + H_2 \rightarrow C_2H_4 + 2H_2 \rightarrow C_2H_2 + 3H_2 \rightarrow 2C_{(solid)} + 4H_2$

Table 1. Summary of Thermochemical Cycles [15].

	Volume Fractions (%)					
	CH_4	C_2H_6	C_3H_8	C_4H_{10}	CO_2	N_2
Methane	100	-	-	-	-	-
Modified Algeria Gas	91.2	6.5	2.1	0.2	-	-
Modified Groningen Gas	83.5	4.7	0.7	0.2	-	10.8
North Sea Gas	88.2	5.4	1.2	0.4	1.4	3.2

Table 2. Compositions of Natural Gas from Different Sources [20].

2.4. Gasification and reforming of coal and biomass

Gasification is a chemical process that converts carbonaceous feedstock into gaseous fuels under a controlled amount of oxygen and/or steam [24]. Main difference between gasification and combustion is that products in gasification have useful heating value. In gasification, pressure inside the gasifier is generally in the range from 20 to 40 bar, whereas methanol or ammonia synthesis requires 50 to 200 bar [25]. In addition, temperatures inside the gasifier is generally in the range from 1400 to 1700°C [25].

Pyrolysis is a thermochemical process that occurs before gasification, and it decomposes the complex hydrocarbons into smaller and less complex molecules in the absence of oxidizers. In pyrolysis, the yield of solar char can be maximized by slowing the heating rate, lowering the temperature or allowing a longer residence time [26]. On the contrary, a higher heating rate, a higher temperature, and a shorter residence time maximize the gas yield. Additionally, liquid yield at an intermediate temperature can be maximized by increasing the heating rate or minimizing the residence time. Tar is an undesired by-product of gasification and pyrolysis. It can cause condensation and consequent plugging, formation of aerosols and polymerization into more complex structures [26].

Gasification is an endothermic process and requires energy to occur. In case of conventional gasification, this energy is supplied from the partial combustion or gasification of feedstock which emits CO_2 to the atmosphere. Use of concentrated solar energy eliminates or reduces the CO_2 emission and utilizes the clean high-temperature gasification process. Additionally, fuel value of the feedstock is increased with solar gasification. For example, fuel value of coal can be increased by about 45% using solar coal gasification [27], and CO_2 emission can be reduced by about 30% when compared to conventional coal gasification [28].

Solar gasification of coal and other carbonecous products is the process of converting these feedstock materials into some synthesis gas (syngas) which includes H_2, CO, CO_2 and water vapor using solar energy [29]. The gasification products can be further processed. For example, syngas can be processed to form methanol or ammonia or used in cement production, and lean gas can be combusted for heating or used in power stations to generate electricity

[16, 30]. Solar gasification can be performed using CO_2 or steam. In general, steam gasification of coal can be written as,

$$Coal + aH_2O \rightarrow bH_2 + cCO + dCO_2 \qquad (8)$$

This process is endothermic and requires temperatures above 1000°C. Solar gasification of petcoke, coal and other carbonecous feedstock started with directly irradiated solar reactors [31]. These designs have high reaction rates and kinetic and high fuel-to-product conversion. However, these reactors have problems with their aperture cover. As a cover, quartz window is commonly used to allow the concentrated solar power into the reactor. In directly irradiated reactors, quartz window has to withstand the high pressures inside the reactor and should not be covered with particles as the gasification occurs. As in the solar cracking, additional flows of inert gases are introduced into the reactor to prevent particle accumulation on the quartz window, but these additional flows introduce additional complexity and cost to the reactor [30]. In EU project SOLSYN [31], a 5-kW reactor prototype was built for solar coal gasification. The temperatures in this reactor could go up to 1700°C with the solar concentrating ratio of nearly 3000, but the general operation temperature was kept at 1220°C. The peak conversion efficiency was found to be 29% [31].

Similar to coal and other carbonecous feedstock, biomass can also be gasified in solar reactors. Conventionally, gasification of biomass has been done using the exhaust gas of combustion of fossil fuels or biomass itself. Biomass includes demol wood, wood chips, sewage sludge, almond shells, straw, etc. If the biomass is used, nearly 30% of the initial biomass has to be combusted with oxygen to drive the gasification process due to the temperature requirement [15]. This temperature requirement varies between 600 – 1000°C [32]. Additionally, one of the other disadvantages of conventional biomass gasification is the formation of tar which blocks and clogs the equipment. There have been some efforts to eliminate the tar formation with proper selection of materials, operating conditions and the design of the gasifier [32].

Solar-assisted gasification of biomass has advantages over the conventional process. The main advantages are the elimination of tar formation, even at temperatures as high as 1200°C, and high and rapid conversion of biomass to syngas. At the National Renewable Energy Laboratory (NREL) of US, bluegrass was gasified with a maximum conversion of 95% and about 5% of the products were hydrocarbons, ash and char [33]. The resident times can be less than 5 seconds [33]. There is also a solar reactor design to combine solar biomass gasification and steam reformation [33].

3. Concentrated solar reactors

This section defines and compares different solar concentrators and gives examples of directly irradiated and indirectly heated solar reactors for the solar fuel production processes defined in the previous section.

3.1. Solar concentrators

There are two main types of concentrated solar collectors, categorized depending on their optical configurations. First type is parabolic trough systems in which there is an absorber tube in the focal line of parabolic reflectors. Linear Fresnel reflectors can also be included in this type of concentrated solar collectors. Second type is point focus solar collectors which include dish systems and heliostats. Dish systems have a solar receiver located in the focal point of the paraboloidal concentrator, and heliostats direct sun light to a solar receiver located at the top of a solar tower. Figure 3 shows the schematic of each solar collector type. Before going into details of each collector type, some terms need to be defined.

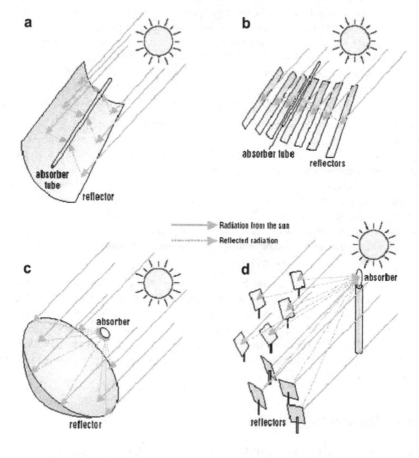

Figure 3. Solar concentrators, (a) parabolic trough, (b) linear Fresnel, (c) dish collectors, and (d) heliostats with solar tower [34].

Concentrating ratio is the ratio of the radiation intensity at the receiver by the radiation intensity received by the concentrator. Concentrating ratio plays an important role on the maximum achievable temperature at the receiver. Solar collector efficiency, η, is a product of Carnot efficiency, η_{Carnot}, and the receiver efficiency, $\eta_{receiver}$, as,

$$\eta = \eta_{Carnot} \cdot \eta_{receiver}$$
$$= \left(1 - \frac{T_o}{T_{rec}}\right)\left(\frac{\alpha \cdot G \cdot C - \varepsilon \cdot \sigma \cdot T_{rec}^4}{G \cdot C}\right) \tag{9}$$

where T_o and T_{rec} are the surroundings and receiver temperatures, respectively, α and ε are the absorptivity and emissivity of the receiver, G is the solar irradiation, C is the concentrating factor and σ is the Stefan-Boltzmann constant (5.67x10^{-8} W/m^{2}K^4). Figure 4 shows the solar collector efficiency as a function of receiver temperature when ambient temperature is 300 K, absorptivity and emissivity are both 1, and the solar irradiation is 1000 W/m^2. The figure shows that higher thermal efficiencies and higher receiver temperatures can be obtained with increasing concentrating ratios. Therefore, the selection of solar concentrators mainly depends on the temperature requirement of the application.

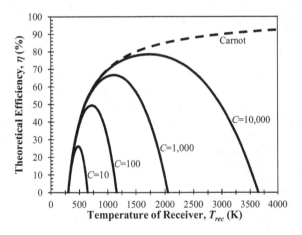

Figure 4. Theoretical solar reactor efficiency, η, as a function of receiver temperature, T_{rec}, for different concentrating ratios, C.

3.1.1. Parabolic-trough systems

Parabolic-trough collectors consist of several tubes interconnected in the focal line of highly reflective sheet material. These linearly connected tubes are generally referred as evacuated tubes since they consist of two concentric tubes whose annulus is vacuumed. The working fluid is circulated in the inner tube. The inner tubes are generally made of metals, and outside of the inner tube has selective coating to increase the absorption and decrease the heat loss. The outer tubes are generally made of glass, and they can also have selective coatings.

Applications of parabolic-trough collectors can be divided into two sections depending on the temperature of application. The low temperature applications, varies from 100°C to 250°C, include domestic hot water, space heating and heat-driven refrigeration [35]. Their concentrating ratios are between 15 and 20. The high temperature applications have temperatures up to 400°C, concentrating ratios of 20 to 30 [35]. Theoretical limit for concentrating ratio can go up to 100 [2]. These collectors are mainly used in power plants that are driven by steam. In the parabolic trough collectors, the pressure within the inner tube can reach 10 MPa.

There are other configurations of cylindrical absorber tubes which are not commonly used [36-38]. One type has a circulation tube inside the inner cylinder which carries the working fluid. This circulation tube is attached to the inner tube with a cylindrical fin. In this type, the absorbing surface is the fin itself, not the inner tube. Another type cylindrical absorber has a delivery tube inside the inner tube, and the working fluid delivered by this delivery tube fills the space in the inner tube. In another tube type, working fluid fills the annulus [37].

These parabolic-trough collectors and other cylindrical absorber tubes may have diffuse or specular reflectors at their back. Diffuse reflectors are generally flat surfaces that cover the entire back of arrays of tubular collectors. However, specular reflectors have parabolic surfaces, and they cover the back of only one cylindrical absorber. If specular reflectors are used, the absorber tubes have to be in the reflector's focal line.

Line focus collectors are mounted with axes either in north-south or east-west direction. Hence, single axis tracking for this kind of collectors is sufficient to track the sun throughout the year. Collectors with axes in north-south direction track the sun from sunrise to sunset each day. Alternatively, collectors with axes in east-west direction track the sun seasonally. The spacing between each line of collectors in a solar farm with parabolic trough collectors is determined considering sun shadow line in the winter when the solar radiation comes with a larger azimuthal angle [38].

As an example, Nevada Solar One is in operation in Boulder City, Nevada, USA since 2007, covers 400 acres and has a capacity of 64 MW [39]. Figure 5 shows a picture of this power plant. Another and largest power plant with parabolic trough collectors is Solar Energy Generating Systems (SEGS) VIII-IX, built on approximately 240 acres and operating at 80 MW each near Harper Lake, California, USA. SEGS are also integrated with conventional natural gas turbines to operate at nights. It was reported that solar energy covers about 90% of the power production [40].

Another design is the absorber tubes with Fresnel reflectors. Fresnel reflectors that are mounted close to the ground direct and concentrate solar irradiation to the absorber tubes that are elevated at a higher level than reflectors. A secondary reflector on the back side of the absorber tube is also use to direct all the irradiation to the absorber tube. The main advantages of using Fresnel reflectors are that they have less wind load than the reflectors of parabolic trough collectors since these reflectors are located at a lower position, and no requirement for vacuum in the absorber tubes and for rotating joints [42]. However, the main disadvantage of Fresnel reflectors is that they have lower concentrating ratios than parabolic trough collectors [43].

Figure 5. Parabolic-trough collectors in Nevada Solar One power plant [41].

In Calasparra, Spain, Novatec Biosol built a power plant with 28 rows of linear Fresnel reflectors that produces 30 MW of electrical power in an area of nearly 200 acres. Figure 6 shows a picture of this power plant. The power plant uses steam, and the temperature and pressure of the steam produced reach to 270°C and 55 bars, respectively [44, 45].

Figure 6. Linear Fresnel reflectors in the power plant Thermosolar Power Plant (PE2) in Spain [46].

3.1.2. Point focus collectors

Parabolic dish collectors concentrate sunlight to the focal point of the parabolic reflectors. These collectors have two-axis sun tracking system. In the focal point of the parabolic reflector, a working fluid is heated directly to a maximum temperature of about 1000 K [47]. This working fluid is generally used to drive a Stirling engine or a gas turbine to produce electricity. The typical parabolic dish collectors have a diameter of 5 to 10 m, and each can produce up to 0.4 MW [47, 48]. The concentrating ratios parabolic dish collectors vary between 1,000 and 10,000 [2]. The reflector is usually made of silver or aluminum coated glass. This kind of collectors can be used in applications with relatively low power requirement in remote areas.

One of the first examples of power plant using parabolic dish collectors was supposed to be Maricopa Solar Plant in Arizona, USA before its contractor company was announced bankruptcy in 2011. Figure 7 shows a picture of this power plant. This power plant consists of 60 parabolic dish collectors that heat the hydrogen to drive Stirling engines. The power plant has a capacity of 1.5 MW. This technology is not commercially operational and available in large scale power production.

Figure 7. Parabolic dish collectors in Arizona, USA [49].

Some solar thermal power plants use arrays of heliostats which are sun-tracking flat mirrors. These mirrors or heliostats with two-axis tracking system direct solar irradiation to the receiver located at the top of a tower on a concrete support. In order to direct the sunlight to the receiver with sufficient accuracy all the times, a motor drive system with a large gear reduction is necessary [50]. Due to the presence of the tower, these power plants are sometimes referred as power tower systems. Concentrating ratios for these systems vary between 500 and 5000 [2], and the temperature at the receiver can exceed 2000 K depending on the concentrating ratio. These power plants can be used for converting solar energy to chemical energy, such applications include reduction of zinc oxide and coal gasification [51].

Throughout the technological development of heliostats, their sizes become larger and larger in order to decrease the production cost since their cost is a strong function of production rate. Although the initial development of heliostats started in 1975, one of the first prototypes of heliostats in 1980s by Sandia Labs, USA had an area of 37 m^2 [52]. Currently, Planta Solar (PS) 20 solar power plants use heliostats with each of their area as 120 m^2 [53]. Another improvement in the development of heliostats is the material choice. Glass mirrors with steel support structure are being replaced by silver polymer mirrors with silver-steel alloy structure in order to increase the structural durability and reduce the weight of heliostats [52]. Some designs also have circular mirrors instead of rectangular ones to reduce stress on the support structure.

The Crescent Dunes Solar Energy Project in Tonopah, Nevada, USA will be an example to the central tower power plants once it is completed late 2013. It will be built on approximately 1600 acres, and it will produce 110 MW of electrical power using molted salt as a phase changing storage medium [54]. As another example, PS 20 which is operational since 2009 has a cavity receiver at the top of a 165-m tower. Figure 8 shows the picture of this power plant. It heats up water in the cavity, and steam reaches an outlet temperature of maximum 550 K. This solar power plant consists of 1255 heliostats with a total area of 30 acres. The solar power plant is backed up with natural gas burnt conventional turbine, and the total power production is 20 MW. PS 20 power tower is cooled with water which is generally replaced by air cooling if the power plant is built on deserts due to lack of water resources. The cooling is necessary for the materials used in the power towers.

Figure 8. PS 10 (back) and PS 20 (front) solar thermal power plants with heliostats with solar towers [55].

To reach the necessary temperatures for the solar fuel production methods given in the previous section, tower or dish type collectors should be used. For further reference, a review paper on the volumetric receivers for the concentrating solar thermal power plants discussed different designs from the projects of the last 3 decades [56]. Another good review was done on comparison of parabolic trough, dish systems, solar towers and tubular systems with Fresnel reflectors by Pavlovic et al. [57].

3.2. Solar reactors

In this section, different solar reactors that were designed for fuel production using concentrated solar energy are discussed and compared in terms of their operating conditions and design parameters. Depending on the reactor design, these reactors were fed feedstock with or without solid particles to perform hydrolysis, cracking, gasification, etc. These particles not only allowed a more uniform temperature distribution inside the reactor, but also helped the reactor to reach higher temperatures faster. These particles also acted as additional reaction sites due to their high surface are to volume ratios. In some designs, feedstock was diluted with some inert gas, such as Argon, to increase the produced fuel yield. An auxiliary gas was also fed to prevent particle deposition on the window surface. In some designs, reactors were supplemented with a cooling system for products to prevent them recombine.

In general, the solar energy is transformed into thermal energy in the structure of volumetric receivers. In some designs, solar energy directly heats the feedstock in the reactor which is referred as directly irradiated solar reactors. Additionally, some of these receivers have a porous metal or ceramic absorber to be heated by solar energy. Metal absorbers can be heated up to 1000°C whereas SiC absorbers can reach 1500°C as maximum temperatures. Then, this thermal energy is transferred to a working fluid that passes through the porous absorber. This kind of reactors is referred as indirectly heated solar reactors.

3.2.1. Directly irradiated solar reactors

In this section, examples of directly irradiated solar reactors are presented with their design parameters, temperature and pressure allowances, their power outputs and their solar fuel production rates. These examples are selected to give a wide range of applications and designs.

Maag et al. [17] tested a 5 kW-prototype of a solar reactor seeded with particles for thermal cracking of methane. The cylindrical reactor was 200 mm in length and 100 mm in diameter. It had a 60 mm aperture area. The concentrator was covered with a 240 mm-diameter quartz window as shown in Figure 9. In their experiments, they used a sun-tracking parabolic concentrator that has a diameter of 8.5 m and could reach a concentrating factor up to 5000 suns. They tested the reactor in the temperature range from 1300 to 1600 K with a concentrating factor of 1720 suns. They varied the volume fraction of carbon in the range of 0 to $7.2x10^{-5}$ and gas inlet flow rate in the range of 8.6 to 15.6 l/min. They reported maximum methane-to-hydrogen conversion of 95% at a residence time less than 2 seconds, and an experimental solar-to-chemical energy conversion efficiency of 16%, whereas their theoretical prediction of the same conversion efficiency was 31%.

Yeheskela and Epstein [58] developed and tested 10-kW particle-seeded solar chemical reactor for producing hydrogen and carbon nanotubes from methane. They used iron pentacarbonyl and ferrocene as catalysts to produce multi-walled carbon nanotubes. The reactor was 300 mm in length, and the quartz window which covered the reactor as shown in Figure 10 was 200 mm in diameter. Additionally, He was used as a screen protector gas to eliminate the particle deposition on and near the window, and N_2 was used as a tornado generator gas. The average reported temperature within the reactor core was 1450°C.

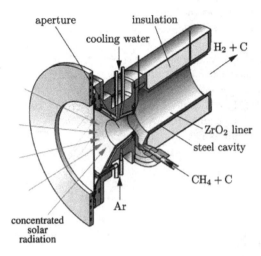

Figure 9. Schematic of design of Maag *et al.* [17].

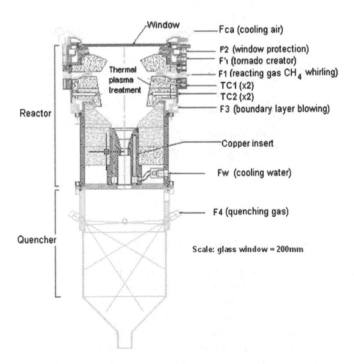

Figure 10. Schematic of design of Yeheskela and Epstein [58].

Abanades and Flamant [59] designed nozzle-type directly-irradiated solar reactor for methane cracking. They used a graphite nozzle with an inner diameter of 10 mm and a length of 65 mm. The schematic of the proposed design is shown in Figure 11. The reactor walls were made of stainless steel, and they were water cooled for their durability. Additionally, the products were also cooled to eliminate recombination of products [60]. The obtained conversion of methane to hydrogen exceeded 95% in molar basis, while the rest of the by-products were C_2H_2, C_2H_4 and C_2H_6. With a direct normal irradiation of 980 W/m², the temperature of graphite nozzle had a maximum temperature of 1385°C, while their model estimated the maximum wall temperature as 1890 K.

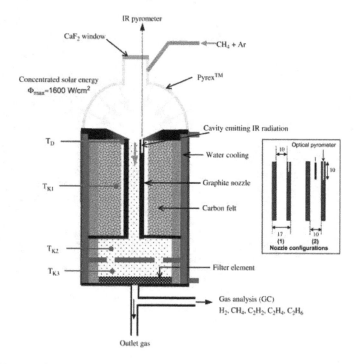

Figure 11. Schematic of design of Abanades and Flamant [59].

Klein et al. [61] investigated the performance of directly irradiated solar methane cracking process with and without CO_2 present in the reactor. The reactor, as shown in Figure 12, had a diameter of 160 mm, a length of 266 mm and an aperture diameter of 60 mm. The gas exit temperatures with CO_2 only (no methane) in the reactor were in the range from 1000 to 1250°C. Additionally, the exit temperatures were in the range from 1100 to 1450°C with CO_2 and methane, when the CO_2 and methane molar ratio were varied from 1:1 to 1:6. Overall, the experimental results were similar to the studies where no CO_2 was introduced into the reactor. Moreover, when the reactor was fed with CO_2 and carbon black and the gas exit temperature reached 1000°C, 20% of carbon particles reacted with CO_2. When the exit gas temperature was increased to 1350°C, about 70% of the carbon particles reacted with CO_2.

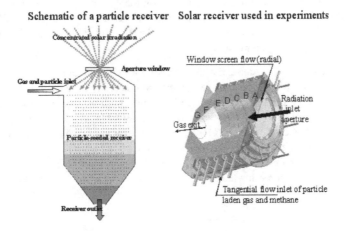

Figure 12. Schematic of design of Klein et al. [61].

Z'Graggen et al. [62] designed a 5-kW prototype reactor for steam-gasification of petroleum coke using concentrated solar energy. The reactor had a 5-cm diameter aperture which was covered by 3-mm-thick quartz window. The window was cooled by oil, and swept by an inert gas to prevent particle accumulation on the window. The solar concentrating ratio was about 5000, and the maximum temperature in the reactor was about 1800 K. The walls of the cavity were covered with Al_2O_3 and insulated from the backside with Al_2O_3-ZrO_2. Figure 13 shows the design of Z'Graggen et al. [62]. As a result of the steam-gasification of petroleum coke, H_2, CO, CO_2 and CH_4 were produced with a chemical conversion ratio of 87%. The overall solar-to-chemical conversion efficiency was about 9%.

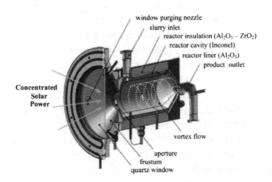

Figure 13. Reactor design of Z'Graggen et al. [62].

3.2.2. Indirectly heated solar reactors

Directly irradiated solar reactors work with high conversion efficiencies. However, they have problems, such as accumulation of particles on the window. In order to overcome this problem and the need for inert gas feeding, indirectly heated solar reactors are alternatively introduced. This section gives some design examples of indirectly heated solar reactors for different solar fuel production methods.

Gordillo and Belghit [63] modeled the reaction kinetics without pyrolysis using finite volume analysis in a two-phase biochar solar gasification reactor with a bubbling fluidized bed as shown in Figure 14. Bubbling was used to introduce fluidizing gases to the reactor. They found that concentrating solar energy and high gas flow rates affected the temperature distribution within the packed bed. Therefore, a uniform temperature distribution could not be obtained which adversely affected the reaction kinetics. Additionally, they showed that although energy conversion efficiency (η), defined as in Equation (10), could be as high as 55%, it decreased with increasing the steam velocity or the bed temperature [64].

$$\eta = \frac{\dot{m}_{product} LHV_{product}}{\dot{Q}_{solar} + \dot{m}_{feedstock} LHV_{feedstock}} \tag{10}$$

where $\dot{m}$ and LHV refer to the mass flow rate and lower heating value, $\dot{Q}_{solar}$ is the solar irradiation, subscripts *product* and *feedstock* denote gaseous products and fed feedstock, respectively.

According to Hathaway et al. [65], problems with the preceding reactor designs had poor heat transfer characteristics [63], formation of ash and tar which block the radiative heat transfer and insulate the reaction zone, and intermittency of solar energy. Hathaway et al. [65] investigated the effects of using molten salt on the reaction kinetics in solar gasification of biomass. For the analysis of pyrolysis which occurs before gasification, they prepared tablets using microcrystalline cellulose, and for the analysis of steam gasification, they used tablets of wood charcoal powder. They carried out a series of experiments in the temperature range from 1100 to 1250 K to show the effects of molten alkali carbonate salts (lithium, sodium and potassium carbonate) on reaction rates using the experimental setup shown in Figure 15. They showed that introducing molten salts increased the rate of pyrolysis by 74% and increased the rate of gasification by more than an order of magnitude since molten salts acted as a heat transfer medium for gasification which ended up with more uniform temperature distribution within the solar reactor. On the contrary to the other studies, the catalytic effect of molten salt on pyrolysis was not observed for the reason that pyrolysis happened rapidly, and then gasification occurred. However, the catalytic effect of molten salt on gasification was observed. Introducing the molten salt increased the pre-exponential factor (i.e., rate of reaction in steady state process) by 24.4 times and increased the activation energy by about 4%. Additionally, using molten salt avoids the tar production (as a by-product of uncatalyzed gasification, occurs especially on startup), and molten salts can act as an energy storage unit to overcome the intermittency effect of solar energy

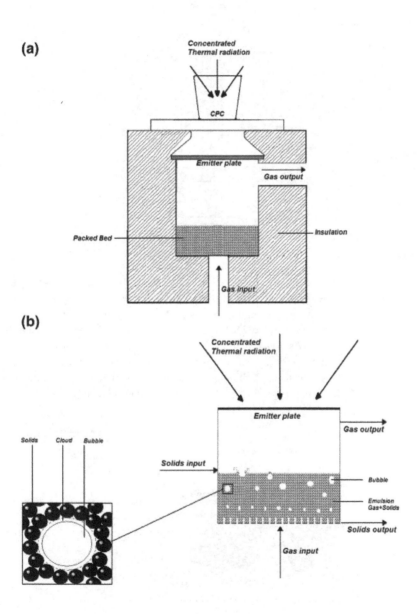

Figure 14. Model of Gordillo and Belghit [63].

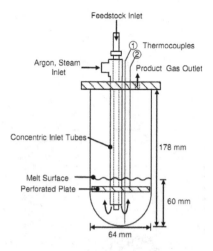

Figure 15. Experimental setup of Hathaway *et al.* [65].

Rodat *et al.* [66] developed a 10 kW tubular reactor prototype, which was indirectly heated, for methane cracking as shown in Figure 16. They used a graphite cubic cavity as a receiver and a quartz window. The quartz window was swept by nitrogen which prevented O_2 to enter the cavity. The reactor reached 2070 K, and the products included C_2H_2 with maximum mole fraction as 7%. As given in Equation (7), C_2H_2 is the last step of H_2 and carbon black evolution. The graphite cavity was purged by N_2. For this configuration, the reactor required about 4000 seconds to reach the required temperature of 1800 K when the experiment started at 300 K under the direct normal irradiance of 1000 W/m² [67].

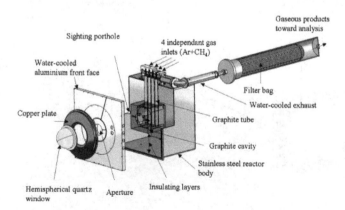

Figure 16. Schematic of design of Rodat *et al.* [66].

Lichty *et al.* [69] designed and analyzed the thermal characteristics of a cavity reactor prototype for solar-thermal biomass gasification as shown in Figure 17. The maximum recorded temperature was 1660 K on the central tube under 7.5 kW power input. They quantified the reacted biomass based on CO and CO_2 as these gases showed the ratio of reactants underwent a complete reaction, and the authors reported an average biomass-to-CO and CO_2 conversion as 58.4%. The residence time was about 4 seconds. They also compared the syngas production of grass and lignin pyrolysis and cellulose gasification using mass spectrophotometer.

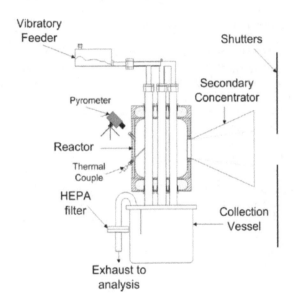

Figure 17. Reactor design of Lichty *et al.* [69].

In the design of the *German Aerospace Center* for directly irradiated solar reactor to reform natural gas is given in Figure 18, porous ceramic absorber coated with Rh catalyst was used [12]. A concave quartz window was also mounted on the concentrating solar collector [12]. The operating conditions were chosen as 1400°C and 3.5 bars, and a volumetric flow rate of 3.8 l/min with 5% methane in argon [70, 71]

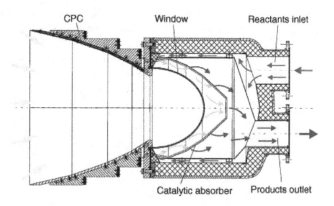

Figure 18. Design of the *German Aerospace Center* for natural gas reforming [12].

Maag *et al.* [72] simulated the performance of a 10 MW commercial-size reactor. The reactor consisted of four graphite absorber tubes with an outer diameter of 24 mm placed in a 0.2 m-cubic graphite cavity as shown in Figure 19. The graphite cavity had an aperture of diameter 9 mm which was covered by quartz window. They predicted 100% methane-to-hydrogen conversion when flow rate of methane was 0.7 kg/s at a reactor exit temperature of 1870 K. Spectral properties of quartz window were estimated using a band model, and view factors were calculated using Monte Carlo ray-tracing method. The energy balance for the overall system was solved with finite volume method. The results showed that it was possible to increase solar-to-chemical energy conversion efficiency from 42% to 60% when the outlet temperature was lowered to 1600 K and, subsequently, the methane flow rate was doubled, but then quality of carbon black as a product would be poorer.

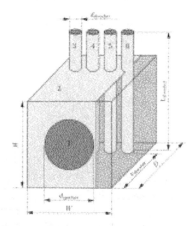

Figure 19. Schematic of design of Maag *et al.* [72].

Osinga *et al.* [73, 74] designed 5-kW indirectly heated solar reactor for the reduction of ZnO. There were two different versions of the reactor. First one had the inner cavity made of graphite, and the second one had the inner cavity made of SiC. Figure 20 shows the second type, reactor with SiC absorber. Both reactors could reach temperature of 1700 K in about 80 minutes after the solar energy was input to the reactor. The reactor with graphite absorber had a vacuum pressure of 10 mbar whereas the pressure inside the reactor with SiC absorber was kept at 1 bar. ZnO and C mixture was reduced to Zn, CO and CO_2 from which Zn can be reacted with water to produce ZnO and H_2 as in Equation (3) [75].

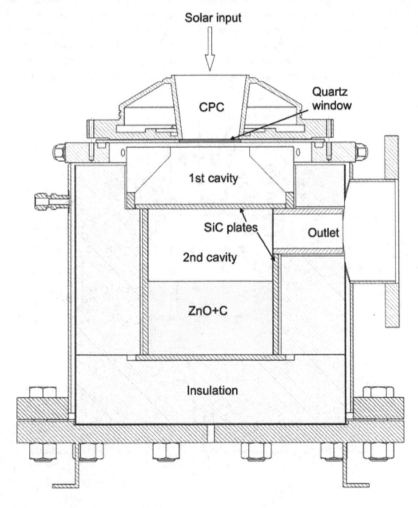

Figure 20. Reactor design of Osinga *et al.* [73].

The reactor with SiC absorber was scaled up to a capacity of 300 kW and operated by Wieckert et al. [76] under the EU project, called SOLZINC. They used solar tower with heliostats to deliver concentrated solar power of 300 kW to the volumetric reactor. The second cavity where the mixture of ZnO and C was present reached to maximum temperature of nearly 1500 K, and overall ZnO-to-Zn conversion reached 95% with a Zn production of 50 kg/h.

Piatkowski et al. [31, 77] designed a 5-kW indirectly heated solar reactor with packed bed for solar steam-gasification. Figure 21 shows the reactor design of Piatkowski et al. The authors used different carbonecous feedstock, such as African coal, sludge and charcoal. Beech charcoal gave the maximum solar-to-chemical conversion efficiency of 29% at a temperature of nearly 1500 K. The solar reactor has two cavities separated an emitter plate made of SiC-coated graphite. The aperture diameter was 6.5 cm which was followed by a 3-mm-thick quartz window. The walls of the lower cavity was also covered by SiC, and insulated by Al_2O_3-SiO_2. The reactor received the concentrated solar energy with concentrating ratios up to 3000, and the maximum temperature measured in the upper cavity was about 1700°C.

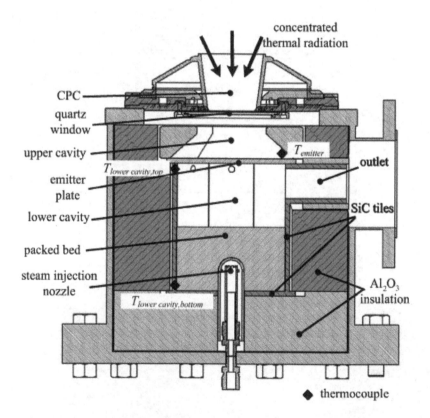

Figure 21. Reactor design of Piatkowski et al. [31, 77]

Summary of the operating conditions of the discussed designs for methane cracking is given in Table 3.

Reference	Maximum Temperature (°C)	Inlet CH_4 Dilution (%vol.)	Reactor Dimensions (mm)	Aperture Diameter (mm)	Inlet Flow Rate(l/min)	Catalytic or Fed Conversion
Directly Irradiated Solar Reactors						
Maag et al. [17]	1327	6-30 (in Argon)	100 (diameter) 200 (length)	60	8.6-15.6	Carbon black seeded
Yeheskela and Epstein [58]	1450	98 (in catalysts)	200 (diameter) 300 (length)	200	5-9.7	Flow with $Fe(CO)_5$, $Fe(C_5H_5)_2$
Abanades and Flamant [59, 60]	1110	11-20(in Argon)	10 (diameter) 65 (length)	10	0.9	No particle feeding
Klein et al. [61]	1471	10-24 (in Argon or CO_2)	160 (diameter) 266 (length)	60	37-60	Carbon black seeded
Indirectly Heated Solar Reactors						
Rodat et al. [66]	1800	10-20 (in Argon)	18 (tube diameter) 200 (cube side)	90	-	No particle feeding
German Aerospace Center [12]	1400	5 (in Argon)	-	-	3.8	Reactor walls with Rh
Maag et al. [72]	1600	10-20 (in Argon)	24 (tube diameter) 200 (cube side)	9	10-48	No particle feeding

Table 3. Operating Conditions of Different Reactor Designs for Methane Cracking.

4. Conclusions and outlook

The research to find an alternative fuel to fossil fuels is led by how the new technologies are economically competitive with the fossil fuel technologies, rather than their efficiencies. However, the economical aspect of fossil fuels should also include the cost for CO_2 emissions or sequestration of CO_2 when fossil fuels are compared to solar fuels since solar fuels have no CO_2 emission. Furthermore, as the fossil fuels deplete and the demand for fossil fuels will exceed their production, their prices will be subjected to significant increase. In this way, the investors in fuel or electricity production sector would see solar fuels as an alternative to fossil fuels with the current level of both technologies. As an outcome, the world would become a more sustainable environment with reduced atmospheric CO_2 level and less pronounced risk for global warming.

It should also be noted that the solar fuel production methods introduced in this chapter are at different levels of maturity. For instance, most of the thermochemical cycles are in laboratory and research scale, whereas gasification and reforming processes are in fully operational or pilot stage. To give some examples on solar fuel production, the targets and predictions of the US Department of Energy (DOE) [78] for both cost and process efficiency are summarized in Table 4 for the ZnO/Zn thermochemical cycle. The predictions show that it is possible and feasible to meet the DOE efficiency and 2015 cost targets. However, the cost target of DOE in 2025 is a challenging objective. The main constituent of cost in thermochemical solar production is the plant capital cost, i.e., cost for heliostats and solar tower, rather than the direct cost for the process. Although process efficiencies are reported and predicted as given in Table 4, the overall solar-to-fuel conversions are still low, less than 10% [1].

	by 2015		by 2025	
	DOE Target	Prediction[+]	DOE Target	Prediction[+]
Cost ($/gge[+])	6	6.07	3	4.18
Process Efficiency (%)	30	35	35	42

[+]gge refers to gallon-of-gas-equivalent.

[+]Based on predicted ZnO-to-Zn conversions of 70% in 2015 and 85% in 2025.

Table 4. Targets of Department of Energy of US [14] and predictions [78] for cost and efficiency for ZnO/Zn thermochemical cycle.

In summary, the advantages of solar fuels include:

- Energy content or heating value of feedstock is increased by converting it to another form, solar fuel.

- Producing storable and transportable fuel which is not possible if solar energy is directly used. Thus, eliminates the intermittency problem of solar energy.

- Solar fuels are clean and sustainable. The thermochemical cycles and thermolysis of water that are used to produce solar fuels have no CO or CO_2 emissions. However, carbon emission occurs for the gasification or reforming of carbonecous feedstocks. If these feedstocks are biomass cultivated with CO_2 from the atmosphere, they are carbon neutral.

On the other hand, these are not mature technologies and still suffer from technical challenges which form the basis for future research including:

- High temperatures needed for solar fuel production processes. High temperatures can be reached with high concentrating ratios. However, high concentrating ratios bring high cost to the system, and high temperatures restrict the material choice

- Recombination of product gases, especially in thermochemical cycles, is a significant problem. This recombination significantly decreases both the process and overall solar-to-fuel efficiency.

- Quenching is introduced to products in order to reduce the recombination. However, quenching adds additional cost and complexity to the reactor and the process management. For some solar thermochemical processes, *membranes* are also required to separate product gases.

- Particle accumulation on the window of the reactor is a problem in directly irradiated solar reactors. This problem can be eliminated by introducing an inert gas with high flow rates to the reactor which further complicates the management of reaction in the reactor. Another solution is to heat the reactor indirectly which reduces the solar-to-fuel efficiency.

- Multiple-step chemical reactions are needed to produce hydrogen in most of the thermochemical cycles. More reactions add further components to the system which increase the cost and the management of the overall fuel production process.

- CO and CO_2 formation can be noteworthy in case of solar gasification and reforming of carbonecous feedstock, although solar fuels are accepted as clean fuels.

These drawbacks of the solar fuel production prevent the technology to be converted to large scale commercially available power plants. However, solar fuel production processes are thermodynamically efficient, favorable developments to increase the feedstock's heating values with the unlimited free solar energy. Therefore, in a long-term prospect, solar fuel production is a promising technology that needs significant research efforts for efficiently producing clean and sustainable fuels.

Author details

Onur Taylan and Halil Berberoglu*

Department of Mechanical Engineering, The University of Texas at Austin, Austin, TX, USA

References

[1] Nowotny J, Sorrell C, Sheppard L, Bak T. Solar-Hydrogen: Environmentally Safe Fuel for the Future. International Journal of Hydrogen Energy. 2005;30(5):521-44.

[2] Steinfeld A, Palumbo R. Solar Thermochemical Process Technology. Encyclopedia of Physical Science and Technology. 2001;15(1):237-56.

[3] Steinfeld A, Meier A. Solar Fuels and Materials. Amsterdam: Elsevier; 2004. p. 623–37.

[4] Baykara S. Experimental Solar Water Thermolysis. International Journal of Hydrogen Energy. 2004;29(14):1459-69.

[5] Steinfeld A. Solar Hydrogen Production via a Two-Step Water-Splitting Thermo-chemical Cycle Based on Zn/ZnO Redox Reactions. International Journal of Hydrogen Energy. 2002;27(6):611-9.

[6] Funke HH, Diaz H, Liang X, Carney CS, Weimer AW, Li P. Hydrogen Generation by Hydrolysis of Zinc Powder Aerosol. International Journal of Hydrogen Energy. 2008;33(4):1127-34.

[7] Abanades S, Charvin P, Lemont F, Flamant G. Novel Two-Step SnO_2/SnO Water-Splitting Cycle for Solar Thermochemical Production of Hydrogen. International Journal of Hydrogen Energy. 2008;33(21):6021-30.

[8] Meier A, Steinfeld A. Solar Thermochemical Production of Fuels. Advances in Science and Technology. 2011;74(1):303-12.

[9] Chueh WC, Haile SM. Ceria as a Thermochemical Reaction Medium for Selectively Generating Syngas or Methane from H_2O and CO_2. ChemSusChem. 2009;2(8):735-9.

[10] Kappauf T, Fletcher EA. Hydrogen and Sulfur from Hydrogen Sulfide—VI. Solar Thermolysis. Energy. 1989;14(8):443-9.

[11] Zaman J, Chakma A. Production of Hydrogen and Sulfur from Hydrogen Sulfide. Fuel processing technology. 1995;41(2):159-98.

[12] Steinfeld A. Solar Thermochemical Production of Hydrogen - a Review. Solar energy. 2005;78(5):603-15.

[13] Harvey WS, Davidson JH, Fletcher EA. Thermolysis of Hydrogen Sulfide in the Temperature Range 1350-1600 K. Industrial & engineering chemistry research. 1998;37(6): 2323-32.

[14] Perret R. Solar Thermochemical Hydrogen Production Research (STCH), Thermochemical Cycle Selection and Investment Priority. Albuquerque, New Mexico and Livermore, California: Sandia National Laboratories 2011. Report No.: SAND2011–3622.

[15] Perkins C, Weimer AW. Solar-Thermal Production of Renewable Hydrogen. AIChE Journal. 2009;55(2):286-93.

[16] Zedtwitz P, Steinfeld A. The Solar Thermal Gasification of Coal—Energy Conversion Efficiency and CO_2 Mitigation Potential. Energy. 2003;28(5):441-56.

[17] Maag G, Zanganeh G, Steinfeld A. Solar Thermal Cracking of Methane in a Particle-Flow Reactor for the Co-Production of Hydrogen and Carbon. International Journal of Hydrogen Energy. 2009;34(18):7676-85.

[18] Hirsch D, Epstein M, Steinfeld A. The Solar Thermal Decarbonization of Natural Gas. International Journal of Hydrogen Energy. 2001;26(10):1023-33.

[19] Ozalp N, Kogan A, Epstein M. Solar Decomposition of Fossil Fuels as an Option for Sustainability. International Journal of Hydrogen Energy. 2009;34(2):710-20.

[20] Rodat S, Abanades S, Coulié J, Flamant G. Kinetic Modelling of Methane Decomposition in a Tubular Solar Reactor. Chemical Engineering Journal. 2009;146(1):120-7.

[21] Abanades S, Flamant G. Hydrogen Production from Solar Thermal Dissociation of Methane in a High-Temperature Fluid-Wall Chemical Reactor. Chemical Engineering and Processing: Process Intensification. 2008;47(3):490-8.

[22] Almodaris M, Khorasani S, Abraham JJ, Ozalp N, editors. Simulation of Solar Thermo-Chemical Hydrogen Production Techniques. ASME/JSME 2011 8th Thermal Engineering Joint Conference; 2011 March 13-17, 2011; Honolulu, Hawaii, USA: ASME.

[23] Rodat S, Abanades S, Flamant G. High-Temperature Solar Methane Dissociation in a Multitubular Cavity-Type Reactor in the Temperature Range 1823 – 2073 K. Energy & fuels. 2009;23(5):2666-74.

[24] Kamka F, Jochmann A, Picard L, editors. Development Status of BGL Gasification. International Freiberg Conference on IGCC & XtL Technologies; 2005; Freiberg, Germany.

[25] Higman C, Burgt Mvd. Gasification. Second ed. Oxford, UK: Gulf Professional Publishing; 2008.

[26] Basu P. Biomass Gasification and Pyrolysis: Practical Design and Theory. Burlington, MA: Academic Press; 2010.

[27] Kodama T, Kondoh Y, Tamagawa T, Funatoh A, Shimizu K, Kitayama Y. Fluidized Bed Coal Gasification with CO_2 under Direct Irradiation with Concentrated Visible Light. Energy & fuels. 2002;16(5):1264-70.

[28] Aoki A, Ohtake H, Shimizu T, Kitayama Y, Kodama T. Reactive Metal-Oxide Redox System for a Two-Step Thermochemical Conversion of Coal and Water to CO and H_2. Energy. 2000;25(3):201-18.

[29] Flechsenhar M, Sasse C. Solar Gasification of Biomass Using Oil Shale and Coal as Candidate Materials. Energy. 1995;20(8):803-10.

[30] Van Heek K. General Aspects and Engineering Principles for Technical Application of Coal Gasification. In: Figuieiredo JL, Moulijn JA, editors. Carbon and Coal Gasification: Science and Technology. Alvor, Portugal: Springer; 1986. p. 383-402.

[31] Piatkowski N, Wieckert C, Steinfeld A. Experimental Investigation of a Packed-Bed Solar Reactor for the Steam-Gasification of Carbonaceous Feedstocks. Fuel processing technology. 2009;90(3):360-6.

[32] Devi L, Ptasinski KJ, Janssen FJJG. A Review of the Primary Measures for Tar Elimination in Biomass Gasification Processes. Biomass and Bioenergy. 2003;24(2):125-40.

[33] Weimer A, Perkins C, Mejic D, Lichty P, inventors; WO Patent WO/2008/027,980, assignee. Rapid Solar-Thermal Conversion of Biomass to Syngas. United States patent WO2008027980. 2008 06.03.2008.

[34] Quaschning V. Technology Fundamentals. Solar Thermal Power Plants. Renew Energy World. 2003;6(6):109-13.

[35] Fernández-García A, Zarza E, Valenzuela L, Pérez M. Parabolic-Trough Solar Collectors and Their Applications. Renewable and Sustainable Energy Reviews. 2010;14(7): 1695-721.

[36] Morrison G, Budihardjo I, Behnia M. Water-in-Glass Evacuated Tube Solar Water Heaters. Solar energy. 2004;76(1):135-40.

[37] Duffie JA, Beckman WA. Solar Engineering of Thermal Processes. 3rd ed. New Jersey: John Wiley & Sons; 2006.

[38] Mills DR, Morrison GL. Compact Linear Fresnel Reflector Solar Thermal Powerplants. Solar energy. 2000;68(3):263-83.

[39] Acciona North America. Nevada Solar One. http://www.acciona-na.com/About-Us/ Our-Projects/U-S-/Nevada-Solar-One (accessed September 9, 2012).

[40] California Energy Commission. Large Solar Energy Projects. http://www.energy.ca.gov/siting/solar/ (accessed September 9, 2012).

[41] Wald ML. The New York Times: In the Desert, Harnessing the Power of the Sun by Capturing Heat Instead of Light. http://www.nytimes.com/2007/07/17/business/ 17thermal.html?_r=1 (accessed September 18, 2012).

[42] Areva. Our Technology and Features. http://www.areva.com/EN/solar-198/arevasolarour-technology.html#tab=tab2 (accessed September 9, 2012).

[43] Abbas R, Montes M, Piera M, Martínez-Val J. Solar Radiation Concentration Features in Linear Fresnel Reflector Arrays. Energy Conversion and Management. 2012;54(1): 133-44.

[44] National Renewable Energy Laboratory. Puerto Errado 2 Thermosolar Power Plant. http://www.nrel.gov/csp/solarpaces/project_detail.cfm/projectID=159 (accessed September 9, 2012).

[45] Novatec Solar. Puerto Errado 2 in Spain. http://www.novatecsolar.com/56-1-PE-2.html (accessed September 9, 2012).

[46] Tubosol PE2. Puerto Errado 2 (in German). http://www.puertoerrado2.com/ (accessed September 18, 2012).

[47] Ummadisingu A, Soni M. Concentrating Solar Power–Technology, Potential and Policy in India. Renewable and Sustainable Energy Reviews. 2011;15(9):5169-75.

[48] Kaygusuz K. Prospect of Concentrating Solar Power in Turkey: The Sustainable Future. Renewable and Sustainable Energy Reviews. 2011;15(1):808-14.

[49] SRP. Tessera Solar and Stirling Energy Systems Unveil World's First Commercial-Scale Suncatcher Plant, Maricopa Solar – Peoria, Arizona. http://energydeals.wordpress.com/2010/06/04/tessera-solar-and-stirling-energy-systems-unveil-worlds-first-commercial-scale-suncatcher-plant-maricopa-solar-with-partner-srp/ (accessed September 18, 2012).

[50] United States Environmental Protection Agency. Final Report: Design and Fabrication of a Reduced Cost Heliostat. http://cfpub.epa.gov/ncer_abstracts/index.cfm/fuseaction/display.abstractDetail/abstract/9033/report/F (accessed September 10, 2012).

[51] Pitz-Paal R, Botero NB, Steinfeld A. Heliostat Field Layout Optimization for High-Temperature Solar Thermochemical Processing. Solar energy. 2011;85(2):334-43.

[52] Jones SA, Lumia R, Davenport R, Thomas RC, Gorman D, Kolb GJ, et al. Heliostat Cost Reduction Study. Albuquerque, New Mexico and Livermore, California.: Sandia National Laboratories 2007. Report No.: SAND2007-3293.

[53] National Renewable Energy Laboratory. Planta Solar 20. http://www.nrel.gov/csp/solarpaces/project_detail.cfm/projectID=39 (accessed September 10, 2012).

[54] Solar Reserve. Crescent Dunes. http://www.solarreserve.com/what-we-do/csp-projects/crescent-dunes/ (accessed September 10, 2012).

[55] Abengoa Solar. PS20, the Largest Solar Power Tower Worldwide. http://www.abengoasolar.com/corp/web/en/nuestras_plantas/plantas_en_operacion/espana/PS20_la_mayor_torre_comercial_del_mundo.html (accessed September 18, 2012).

[56] Ávila-Marín AL. Volumetric Receivers in Solar Thermal Power Plants with Central Receiver System Technology: A Review. Solar energy. 2011;85(5):891-910.

[57] Pavlović TM, Radonjić IS, Milosavljević DD, Pantić LS. A Review of Concentrating Solar Power Plants in the World and Their Potential Use in Serbia. Renewable and Sustainable Energy Reviews. 2012;16(6):3891-902.

[58] Yeheskel J, Epstein M. Thermolysis of Methane in a Solar Reactor for Mass-Production of Hydrogen and Carbon Nano-Materials. Carbon. 2011;49(14):4695-703.

[59] Abanades S, Flamant G. Experimental Study and Modeling of a High-Temperature Solar Chemical Reactor for Hydrogen Production from Methane Cracking. International Journal of Hydrogen Energy. 2007;32(10-11):1508-15.

[60] Abanades S, Flamant G. Production of Hydrogen by Thermal Methane Splitting in a Nozzle-Type Laboratory-Scale Solar Reactor. International Journal of Hydrogen Energy. 2005;30(8):843-53.

[61] Klein HH, Karni J, Rubin R. Dry Methane Reforming without a Metal Catalyst in a Directly Irradiated Solar Particle Reactor. Journal of solar energy engineering. 2009;131(2):021001.

[62] Z'graggen A, Haueter P, Trommer D, Romero M, De Jesus J, Steinfeld A. Hydrogen Production by Steam-Gasification of Petroleum Coke Using Concentrated Solar Power – II Reactor Design, Testing, and Modeling. International Journal of Hydrogen Energy. 2006;31(6):797-811.

[63] Gordillo E, Belghit A. A Bubbling Fluidized Bed Solar Reactor Model of Biomass Char High Temperature Steam-Only Gasification. Fuel processing technology. 2010;92(3):314-21.

[64] Gordillo E, Belghit A. A Downdraft High Temperature Steam-Only Solar Gasifier of Biomass Char: A Modelling Study. Biomass and Bioenergy. 2011;35(5):2034-43.

[65] Hathaway BJ, Davidson JH, Kittelson DB. Solar Gasification of Biomass: Kinetics of Pyrolysis and Steam Gasification in Molten Salt. Journal of solar energy engineering. 2011;133(2):021011.

[66] Rodat S, Abanades S, Flamant G. Co-Production of Hydrogen and Carbon Black from Solar Thermal Methane Splitting in a Tubular Reactor Prototype. Solar energy. 2011;85(4):645-52.

[67] Rodat S, Abanades S, Sans JL, Flamant G. Hydrogen Production from Solar Thermal Dissociation of Natural Gas: Development of a 10 kW Solar Chemical Reactor Prototype. Solar energy. 2009;83(9):1599-610.

[68] Rodat S, Abanades S, Flamant G, editors. Hydrogen Production from Natural Gas Thermal Cracking: Design and Test of a Pilot-Scale Solar Chemical Reactor. 18th World Hydrogen Energy Conference 2010; 2010 May 16-21, 2010; Essen, Germany.

[69] Lichty P, Perkins C, Woodruff B, Bingham C, Weimer A. Rapid High Temperature Solar Thermal Biomass Gasification in a Prototype Cavity Reactor. Journal of solar energy engineering. 2010;132(1):011012.

[70] Wullenkord M, Funken KH, Sattler C, Pitz-Paal R, Stolten D, Grube T, editors. Hydrogen Production by Thermal Cracking of Methane–Investigation of Reaction Conditions. 18th World Hydrogen Energy Conference; 2010; Essen, Germany.

[71] Worner A, Tamme R. CO2 Reforming of Methane in a Solar Driven Volumetric Receiver-Reactor. Catalysis today. 1998;46(2-3):165-74.

[72] Maag G, Rodat S, Flamant G, Steinfeld A. Heat Transfer Model and Scale-up of an Entrained-Flow Solar Reactor for the Thermal Decomposition of Methane. International Journal of Hydrogen Energy. 2010;35(24):13232-41.

[73] Osinga T, Olalde G, Steinfeld A. Solar Carbothermal Reduction of ZnO: Shrinking Packed-Bed Reactor Modeling and Experimental Validation. Industrial & engineering chemistry research. 2004;43(25):7981-8.

[74] Osinga T, Frommherz U, Steinfeld A, Wieckert C. Experimental Investigation of the Solar Carbothermic Reduction of ZnO Using a Two-Cavity Solar Reactor. Journal of solar energy engineering. 2004;126(1):633-7

[75] Epstein M, Olalde G, Santén S, Steinfeld A, Wieckert C. Towards the Industrial Solar Carbothermal Production of Zinc. Journal of Solar Energy Engineering. 2008;130(1): 014505.

[76] Wieckert C, Frommherz U, Kräupl S, Guillot E, Olalde G, Epstein M, et al. A 300 Kw Solar Chemical Pilot Plant for the Carbothermic Production of Zinc. Journal of Solar Energy Engineering. 2007;129(2):190-6.

[77] Piatkowski N, Steinfeld A. Solar-Driven Coal Gasification in a Thermally Irradiated Packed-Bed Reactor. Energy & Fuels. 2008;22(3):2043-52.

[78] Kromer M, Roth K, Takata R, Chin P. Support for Cost Analyses on Solar-Driven High Temperature Thermochemical Water-Splitting Cycles. Lexington, MA: TIAX, LLC2011 February 22, 2011. Report No.: DE-DT0000951.

Chapter 2

Sustainability in Solar Thermal Power Plants

Rafael Almanza and Iván Martínez

Additional information is available at the end of the chapter

1. Introduction

In the last two decades, there has been growing interest in developing indicators to measure sustainability, which is currently seen as a delicate balance between the economic, environmental and social health of a community, a nation or even our planet. The current measure of sustainability tends to be an amalgam of economic, social and environmental aspects. Economic indicators have been used to measure the state of regional economies for over a century, and social indicators are largely a phenomenon of the postwar world. However, environmental indicators are relatively new and attempt to incorporate the ecosystem into the socio-economic indicators of a study site.

Any interest in defining these indicators primarily comes from the need to monitor performance and to indicate improvements resulting from specific actions. While economists have little difficulty in applying quantitative indicators, sociologists can have great difficulty in creating useful indicators for assessing the quality of life of a social group, as this is an issue that can be approached from different perspectives, many of them intangible. Scientists involved with the environment are considered less likely to have difficulty in establishing practical indicators with which to assess the ecological integrity of an ecosystem, either generally or in specific qualifying aspects.

However, sustainability is more than the interconnection of the economy, society and the environment. It may be something greater and more noble than a dynamic, collective state of grace, a theory such as Gaia (a set of scientific models of the biosphere in which life is postulated that fosters and maintains suitable conditions for itself, affecting the environment), or even the spirit. Instead of asking how can we measure sustainability, it may be more appropriate to ask what degree of sustainability is it?

1.1. The concept of sustainability

The concept of sustainability has penetrated most life spheres, not only as a political requirement but also as something that clearly resonates deep within us, even if we have a poor understanding of what it is. The concept first emerged in the mid-1970s, but it exploded on the world stage in 1987 with the Brundtland Report (1987), in which sustainable development was defined as meeting present needs without compromising the ability of future generations to meet their own needs.

Even though this is a very noble goal, this definition challenges interpretation or operational implementation. Most of us would see our personal needs in the context of our circumstances and not as absolute entities. Therefore, our perception of the needs of future generations would impoverish the imagination. "How much is enough?" is a question we have to explore together, but it can only be answered separately. However, we rarely ask this key question individually, let alone collectively.

Once the Earth's ecological integrity is assured and our basic needs are met, how much is enough? The question should be considered in most developed countries, where in the midst of wealth there is still inequality. Increasing inequality is a necessary characteristic for the growth and advancement of an economy. Although it is desirable to achieve a high standard of living, there are finite limits. Our concern for the environment generally decreases with more prosperity, and we should not expect that our pursuit of sustainability should increase as our material wealth increases. Kerala, Cuba, Mennonite and Amish communities are all examples of small societies that practice sustainability, and they all exhibit traits of greater equity, justice and social cohesion.

There are other definitions that ignore human needs and express sustainability in terms of ecological integrity, diversity and limits. However, these definitions also challenge objective interpretation. Such deficiencies in the definitions can cause considerable frustration in a rational way of thinking, particularly for those trying to measure sustainability (Trzyna, 1995). Meanwhile, a reductionist mindset has the ability to link quantitative and productive activity, as in the case of sustainable agriculture, forestry, land management, fisheries, etc. Consequently, growth and sustainable development have been captured as the dominant paradigm. Sustainable development is held up as a new standard for those who really do not want to change the current model of development (Gligo, 1995), and sustainable development alone does not lead to sustainability. In fact, it is possible to support the longevity of an unsustainable path (Yanarella and Levine, 1992). However, the concept is still with us and is becoming stronger.

In general, we have a better understanding of what is unsustainable rather than what is sustainable. Unsustainability is commonly seen as the degradation of the environment (in its broadest sense), the strains of the human population, wealth and green technology in its global limits. Because these effects are entirely of our own construction, their control is, at least in theory, within our capabilities. Human nature tends to promote physical and biological limits towards survival rather than sustainability. We likely think of sustainability in terms of justice, interdependence, sufficiency, choice and above all, (if we were to think deeply about it) the meaning of life.

Sustainability is also non-material life–the intuitive, emotional, and spiritual creativity for those who strive for all forms of learning. Perhaps some truths are really fundamental and universal, if their meaning and spirituality are components of sustainability. These morals and values, however, are not necessarily absolute and can be very difficult to define. For example, values are qualities that are derived from our experiences. If they confirm our default values, then we are more likely to adopt these values. When our experiences are continually at odds with implicit values, we are more prone to change our personal values with respect to the projected values.

Our inability to define sustainability means that we cannot prescribe it. The future may develop according to our vision and ability to always recognize global limits. Sachs (1996) presents three perspectives of sustainable development: the competition implies the prospect that infinite growth is possible over time; the astronaut perspective recognizes that development is poor over time; and the home accepts the prospect of finitude in development. These may be respectively considered as the dominant paradigm perspective, the precautionary principle and conservationist.

Accepting sustainability as a concept can create as many difficulties as the concept of evolution did 150 years ago. During this time, we have not addressed physical consequences involving the collective proficiency requirements for all companies; thus, in general, human awareness has created the concept of ecological crisis with little consequence. Therefore, any discussion of sustainability is essentially a debate about the meaning of what, who, why and how we believe individually and collectively. However, we are very reluctant to participate in the debate on a collective basis, even locally, let alone nationally or globally, in part because it is a messy and time-consuming proposition, i.e., there is a crisis of perception on which one side resides banality, while on the other side there is uncertainty and fear.

1.2. General indicators

Indicators and measurements are essential components in closed physical systems because they are an integral part of the scientific method. In this context, each indicator must be enclosed between target value limits to guide political and social action. Its usefulness for socio-biophysical closed systems (e.g., human welfare) and, in particular, to open physical systems (e.g., businesses, national economies, regional sustainability) is still unknown because knowledge of the full impact of external factors may not be possible. However, the Earth is ultimately a closed system, except for the flow of energy. In that sense, measures are needed that are theoretically possible globally, but local measures are potentially more meaningful and actionable. The impact of some issues can only be evident at the global level, for example, global warming and ozone depletion, even though the solutions may be local.

Henderson (1991) wrote extensively on indicators, and particularly on current paradigms. The proliferation of indicators is indicative of the confusion and uncertainty of what has been measured as well as the absence of debate and understanding

1.2.1. Economic indicators

There is much discontent with economic indicators, as the majority states that they are indicators of something more than the economy. Some do not believe that there are significant economic measures of sustainability.

The most common indicator is the Gross National Product (GNP), now replaced by the Gross Domestic Product (GDP). Daly and Cobb (1994) developed the Index of Sustainable Economic Welfare (ISEW), which recently has been refined as the Genuine Progress Indicator (GPI) by Cobb, et al. (1995). Consumption remains the basis of the index, but instead of adding only negative or harmful consumption (e.g., environmental protection), positive beneficial consumption is also added (for example, voluntary work, child care, housework, etc.). It is difficult to conceive of an index where consumption is the basis for measuring sustainability.

GDP and GPI are aggregations of specific economic indicators, which can be equally sensitive in terms of time, or the actions of adjustment, but which do not apply to social or environmental concerns. Economic indicators are therefore not particularly useful as measures of sustainability, even though economic considerations must be taken into account.

The basis of modern economic theory has a political and a cultural component that addresses scarcity of resources. Affirming the need for a theory that goes beyond that and reflects the basic human needs would be very helpful.

1.2.2. Social indicators

Overall, there are five types of social indicators: informative, predictive, problem-oriented, program evaluation, and goal set. Several of them are partly economic, environmental and sustainable; they can be combined and compared, such as socio-economic indicators.

Indicators such as the standard of living, which is measured by analyzing time series data on observable phenomena, are called objective. Indicators such as quality of life, which measures the perceptions, feelings and responses through questionnaires with classified scales, are called subjective. The correlation between these conditions is very low, and there are considerable difficulties related to indicator aggregation and the design of weighting schemes.

1.2.3. Environmental / ecological indicators

Environmental indicators tend to relate most closely to human activity but may include economic, social and sustainability parameters. Measures may include the quality of living conditions and work environments, including air, land and water, as well as the productive use of resources.

Ecological indicators are more concerned with natural ecosystems; in some cases, human impact is not as obvious. The indicators for the integrity of ecosystems and biodiversity are prominent. The OECD produces a "pressure-state-response" model that many countries have used in the preparation of their "State of the Environment."

Most indicators have thresholds and targets. At present, there seems to be no drive to aggregate indicators or obtain a unique index. However, the Framework Convention on Climate Change (UNFCCC) and the Global Environment Facility (GEF) have very specific defined indicators for the problem of global climate change that may be adopted by all countries. Table 1 provides an overview of each topic in the UNFCCC so that each ecosystem could be described as the physical state of the substances found therein. As indicators for the GEF, they have a direct relationship with the strategic objectives defined by the same agency for action.

Topic	Sub-topic	Indicator	Unit of measurement
Atmosphere	Climate change	Emission of greenhouse gases	Gg or ton of CO_{2eq}
	Decreased ozone	Consumption of substances that deplete the ozone layer	Ton of CFC-11 or CFC-12 equivalent
	Air quality	Concentration of air pollutants in urban areas	$\mu g/m^3$, ppm, ‰
Land Use	Agriculture	Area of arable land under cultivation and permanent	ha
		Fertilizers	kg/m^2
		Pesticides	kg/m^2
	Forest	Forest area as a percentage of the total area	%
		Intensity of logging	%
	Desertification	Area affected by desertification	km^2 or %
	Urbanization	Area occupied by informal and formal settlements	km^2
Oceans, seas and coasts	Coastal areas	Concentration of algae in coastal waters	mg of chlorophyll/m^3
		Percentage of the total population that lives in coastal areas	%
	Fishing	Annual catch of target species	Ton/year
Fresh water	Water distribution	Annual withdrawals of ground and surface water as a percentage of total available water	%
	Water Quality	Biochemical Oxygen Demand (BOD) in water bodies	mg/L
		Concentration of fecal coliform in freshwater	mg/L
Biodiversity	Ecosystem	Area covering selected key ecosystems	km^2 o ha
		Protected area as a percent of total land area	%
	Species	Abundance of selected key species	# of individuals

Table 1. Outline of indicators proposed by the UNFCCC for global climate change.

1.2.4. Indicators of sustainability

Sustainability measures today tend to be an amalgam of economic, environmental and social indicators. The first two are susceptible to quantitative measurement because they can be expressed in biophysical terms, while the third is not easily quantified. Therefore, there is a tendency to only view biophysical sustainability.

Examples of sustainability indicators for a city are as follows:

• Per capita income.

• Solid waste generated/water consumption/energy consumption per capita.

• Proportion of workforce at the ten largest employers.

• Number of days of good air quality per year.

• Diversity and population size of particular urban wildlife (especially birds).

• Distance traveled on public transport and private transport per inhabitant.

• Residential density in relation to public space in city centers.

• Hospital admissions for certain types of childhood diseases.

• Percentage of children born with low birth weight.

Boswell (1995) proposed a theoretical basis for sustainable development indicators on a foundation of knowledge in sociology and ecology. Below, we present a set of attributes (energy use, community structure, life history, nutrient cycling, selection pressure and balance) in terms of objectives for the sustainable management of communities. The system lists 23 necessary conditions, but this may not be sufficient. The same author evaluates these goals with the selected sustainable development indicators. While a human ecology approach is clearly appropriate, Boswell (1995) does not recognize that the communities themselves should determine the strategy and indicators.

Whereas these are facets of sustainability, we must look beyond conventional measures to include a sense of quality of life, welfare, relevance, and harmony. We may have to be willing to accept semi-quantitative and qualitative indicators.

Environmental and social indicators are rarely expressed with a unique index. There is some interest in developing a single sustainability index based on a weighting of economic, environmental and social criteria, but this index cannot meet response times ranging from a few years (e.g., medical intervention) to a generation (e.g., global warming).

1.3. Criteria for the selection of sustainability indicators

The monitoring of sustainability is a long-term exercise so it must be flexible. The criteria for selecting appropriate indicators today could be expressed in a straight line with a slope and perhaps a long learning curve, and our ideas and preferences may change over time once complex criteria can achieve amenable results through statistical analysis. Perhaps someon can reduce a large set of indicators into a single sustainability index. Conversely, some communities may prefer or be willing to accept a few qualitative indicators for the sake of simplicity and direct relevance. Excluding qualitative criteria because they are not easily amenable to objective analysis would likely lead to the exclusion of essential characteristics of sustainability.

The numerous sets of criteria, e.g., Liverman (1988) and Seattle (1998), range from the simple (efficiency, fairness, integrity, management skills) to the complex. Hart (1995) believes that the best measures are not yet developed but suggests the following criteria:

- Multi-dimensional, linking two or more categories (e.g., economy and environment).
- Looking to the Future (range 20 to 50 years).
- Emphasis on local wealth, local resources and local needs.
- Emphasis on the levels and types of consumption.
- Measures and visualize changes that are easy to understand.
- Reliable, accurate and updated data available.
- Reflect local sustainability to improve global sustainability.

Social criteria (e.g., quality of life, sense of security, relationship with others) must reflect the degree of choice that a person has in an action. Many of us are locked into our own systems of collective construction within the dominant paradigm (many unsustainable) where the choice of being different can be socially, economically and practically difficult. Examples of this are the use of solar radiation and precipitation in dwellings and foregoing ownership of a car.

1.4. Risk analysis and comparative risk assessment

In all stages of information, including insufficient quality and quantity or vagueness and uncertainty, where much is at stake and there are several options for action, risk analysis can assist in selecting the most accurate values, lower costs, and/or the lower-risk option. The poorer the information, the greater the uncertainty, and risk analysis may be necessary. We suggest a preliminary stage of data analysis in order to confront a different set of issues and problems with inadequate resources. This technique classifies the problem issues according to the urgency, cost and likelihood of success.

It is frequently argued that there is insufficient or inadequate information to permit taking a rational action, including activities that affect sustainability. However, we know that there are systemic functional weaknesses in both ourselves and in our organizations. Research information actually adds to the uncertainty or controversy; we lose valuable time while more unnecessary work is undertaken. We know the direction that our action should take, but we do not know exactly what that action should be. Many of the problems and solutions are neither technically nor entirely rational. A new methodology for needs that arise may be required for sustainability. They should only be started through social action, where the general population as well as technical experts report on issues and decision-making recommendations.

1.5. Limitations of the measures of sustainability

Although we cannot objectively and unambiguously define sustainability, we must not abandon or postpone attempts to measure it. Even if we recognize that there are other equally valid ways of learning, we must begin where we are, even though that may be reductionist, rational or materialistic. We can define the limiting aspects (for example, the sustainable productive capacity of a specific area of the Earth) and trends in the direction of sustainability (for example, increased use of public transport, a more equitable distribution of revenues) and choose indicators that are appropriate and meaningful. The former must be below the threshold of unsustainability. The latter must give directions that require us to act. Many, in fact, are actually indicators of unsustainability. Many discussions and studies on the measurement of sustainability are not defined, nor do they even provide a common understanding of what is measured. The context of sustainability cannot be separated from the measurement.

We recognize at the outset the limitations of quantitative measures. However, we must be on guard to keep the threshold clear. Although sustainability is about quality and other intangible non-physical aspects of life, this does not mean that we are unable to obtain measurements for them. Just as biological indicators (e.g., health of trouts) are now used to measure the quality of industrial effluents alongside conventional physical-chemical indicators, we must be able to obtain parameters that serve us and the Earth.

1.6. Some indicators to measure sustainability

If we know that we are becoming more sustainable without having to measure the "sustainability discourse" as part of the process that then leads to a sustainable lifestyle and measures of it, some of which are relatively easy to measure and some of which are roughly quantified to preset limits. However, if it is consistent, then we can say that achieving sustainability has begun. Therein lay the success of initiatives such as Seattle.

The initial challenge of this discourse is communicating the environmental and social change that is underway within organizations, as groups cannot yet see their particular success as part of the combined progress towards sustainability. The dialogue should be extended to the wider community to open the discussion for a more effective participation on the big issues ahead. Local communities need to renegotiate their sense of community in the modern world and discover new modes of expression.

2. Evaluation of sustainability between combined cycle power plant production and a hybrid solar-combined cycle system

A discussion on sustainable development must create a process-oriented dialogue and therefore a dynamic concept that establishes priorities; the generic concept of sustainable development must be able to determine its specificity and concreteness at a local and regional level

This section presents a biophysical, social and economic need for high available renewable resources within our country. As a primary energy source for the generation of electricity in a combined cycle solar energy plant, the sustainability is measured in physical terms, while also taking social, economic and environmental interactions into account.

2.1. Overview

We consider the following questions: What is to be held, for how long, and at what spatial scale? These questions involve social concepts and economic and biophysical factors that should be evaluated as deeply as the scope of this study can allow. This project should also be evaluated superficially, viewing sustainability as a multivariate feature in a socio-environmental system that involves answering additional questions such as: Sustainability for whom, who will carry it out, and how can it be done? Only then can we understand and integrate the plurality of preferences, priorities, perceptions and joint inequalities in the objectives of what is to be held in an appropriate application to the different scales of analysis.

This section requires the evaluation of sustainability between two electrical generation systems: a combined cycle (conventional) and a hybrid solar-combined cycle.

Energy is essential for economic, social and global welfare, but unfortunately, most of it is produced and consumed in unsustainable ways (Yuksel, 2008). The primary source is fossil fuels (oil, coal and natural gas), with more than 90% of global production used to meet commercial energy needs. OPEC forecasts foresee further growth into 2030, both in developed and developing countries. Consequently, energy poverty is a crucial variable for the foreseeable future (OPEC, 2007), while the control of gases and other substances emitted into the atmosphere will become a more urgent matter to be resolved. This condition implies that further improvements must be achieved in the production, transmission, distribution and consumption of electricity (Yuksel, 2008).

In this context, renewable energy sources such as solar, wind, hydro, geothermal and biogas are potential candidates to meet global energy requirements in a sustainable manner. Renewable energy sources have some advantages when compared with fossil fuels (Demirbas, 2000). As a result, the increased use of renewable energy can have a significant environmental effect.

Among renewable sources, solar technologies are attracting worldwide attention (Patlitzianas et al., 2005, Hang et al., 2007); their application in new structures and their adoption in existing ones is currently one of the more common approaches with respect to electricity and heating supply. For example, solar photovoltaic technology worldwide in 2004 reached a production level of 1256 MWp, a 67% increase in production from 2003 (Flamant et al., 2006). Photothermal technology has reached over 430 MW (Morse, 2008).

This trend is expected to continue in the coming years, requiring the creation of specific tools for evaluating the efficiency of solar technology. Classical methods essentially provide tools for the assessment of energy and economy. However, placing these assessments in the broader context of sustainability of the environment require more integrated analyses. From this perspective, one must quantify both environmental and economic costs

To this end, "emergy" has recently been identified as a valid analysis approach. Emergy can be defined as "useful energy (exergy) of a certain type, which has been used both directly and indirectly in the process of developing a particular product or service" (Odum, 1988; Scienceman and El-Youssef, 1993). Emergy expresses the cost of process-equivalent units of energy, such as solar power. The basic idea is that solar energy becomes the primary unit of energy that expresses the value of any other unit of energy so that it is possible to compare completely different systems, such as Emjoule (emergy joule), also called the emjoule solar, which is designated by the symbol (sej). "Emergy calculations have the same purpose as the Exergy: to capture the energy hidden in the organization and construction of living organisms." It is beyond words to define emerging as "exergy built."

2.2. Definitions

The means used to achieve the desired objectives will be varied, so emphasis should be placed on long-term ecological sustainability. All methods should promote the efficient use of energy and resources, encourage the use of renewable energy sources (and thereby reduce fossil fuel use), reduce costs and increase the efficiency and economic viability of alternative energy sources.

From the environmental point of view, the sustainability of a hybrid solar-combined cycle power generation system will essentially depend on the management and optimization of the following processes:

- Reduction in natural gas consumption; this will also reduce greenhouse gas emissions into the atmosphere.

- Preservation and integration of biodiversity; the use of parabolic trough concentrators requires a large area, so it is important to locate the hub area while affecting as little regional flora and fauna as possible.

Socially, electric power generation should benefit all communities in the area no matter where they are located, thus providing people with energy that can be used to increase productive capacities, self-management and local cooperative mechanisms. One can say that the process is a socially driven activator, improving conditions for all those who receive that energy as well as future generations.

2.3. Systemic attributes and operational definitions of a sustainable management system

The following primary schematic characteristics of a sustainable system must be analyzed:

- Productivity: the system's ability to provide the required level of goods and services. Represents the attribute value over a period of time.

- Equity: the system's ability to deliver productivity (benefits and costs) in a fair manner. This implies a distribution of productivity among affected beneficiaries in the present and the future

- Stability: refers to ownership of the system having a dynamic state of equilibrium. It can maintain the productivity of the system at a level not decreasing over time under normal conditions.

- Resilience: the ability to return to equilibrium or maintain productive potential after the system has suffered major disturbances.

- Reliability: the ability of the system to be maintained at levels close to the usual equilibrium, i.e., temperature shocks.

- Adaptability and flexibility: the ability to find new equilibrium levels to long-term changes in the environment. It is also the ability to actively seek new levels of productivity.

- Self-reliance or self-management: the ability to regulate and control the system through outside interactions. Includes organizational processes and mechanisms of socio-environmental systems to endogenously define their own goals, priorities, identities and values.

It also emphasizes that the sustainability of a system depends on endogenous properties and their external linkages with other systems and structural relationships. These attributes are designed to apply to systems management as a whole, including social, economic, environmental and technological attributes. Focusing on the abovementioned attributes allows for the development of sustainability indicators fundamental to systemic priorities, thereby avoiding long lists of purely descriptive factors and variables.

In operational terms, a sustainable management system will be one that simultaneously allows the following:

- A high level of productivity through efficient and synergistic use of natural and economic resources.

- Reliable production, stable and resilient to major disturbances in the course of time, ensuring access and availability of productive renewable resources; the use, restoration and protection of local resources, proper temporal and spatial diversity of the natural environment and economic activities and risk-sharing mechanisms.

- Adaptability or flexibility to adjust to new conditions of economic and biophysical environment through innovation and learning processes and the use of multiple options.

- Fair and equitable distribution of costs and benefits to the different affected groups, ensuring economic access and cultural acceptance of the proposed systems.

- An acceptable level of self-reliance to respond and externally manage induced changes, maintaining its identity and values.

These five general attributes of sustainability are the basis for the design of indicators:

- Productivity, which can be evaluated by measuring efficiency, achieved average returns and availability of resources.

- Stability, reliability and resilience, which can be evaluated with the trend and variation in the average return, with the quality, conservation and protection of resources, renewability in the use of resources, spatial and temporal diversity systems with a relationship between income and opportunity cost system, and an evolution of jobs created and risk-sharing mechanisms.

- Adaptability, which can assess the range of technically and economically available options, with the ability to change and innovate, strengthening the relationship between the process of learning and training.

- Equity, which can assess the distribution of costs and benefits to participants and target groups, and the degree of "democratization" in the decision-making process.

- Self-reliance, where one can evaluate the forms of participation, organization and control over the system and decision-making.

This project follows the methodology proposed by Masera (1996), which consists of the following:

1. Determining the objectives of the evaluation and defining the management systems to assess their characteristics and the socio-environmental assessment.

2. Selecting the indicators that define the critical points for the sustainability of the system, the diagnostic criteria and the derived sustainability indicators.

3. Measuring and monitoring indicators, including the design of analytical instruments and the procedure used to obtain the desired information.

4. Obtaining and submitting results that compare the sustainability of the analyzed management systems, identify the main obstacles to sustainability, and provide suggestions for improving the system of innovative management.

2.4. Objective of the assessment: Definition of management system

Planning to meet the nation's future electricity demand is an issue of paramount importance, considering the urgent need for economic development, the projected population growth and the allocation of capital to finance that growth.

The Mexican Federal Electricity Commission (CFE) forecasts that the demand for electricity will grow 5 to 6% annually in coming years, requiring a dramatic increase in production capacity along with new schematic development. A primary objective for the electricity sector should be a transition from a centralized power system to a geographically distributed and decentralized system, allowing for a wide availability of natural resources in order to focus on the use of renewable energy. In this scheme, we propose a reduction in plant size and geographical dispersion.

In this paper, we carry out a sustainability assessment by comparing a traditional production of electricity through a 316 MW combined cycle plant using natural gas as the primary energy source as well as an innovative system of electricity production through a solar-hybrid combined cycle plant that employs an 80 MW thermal and a 236 MW natural gas energy source.

The large capacity 80 MW solar plant was chosen because of economy of scale (a larger capacity involves occupying a land area of over one million square meters). This capacity is not arbitrary; it is based on an example presented by PURPA (Public Utility Regulatory Policies Act) in the United States of America, which arose from limits for small producers. We selected this capacity to generate the parabolic trough based on the experience and cost information at our disposal. ABB GT24 combined cycle turbines of 236 MW and 316 MW were selected for use in the different alternatives.

The following are the main determinants used to characterize the proposed systems:

2.4.1. Bio-physical components of the system

For solar thermal technologies to be effective, the proposed systems must be located in an area with high solar irradiance during most of the year. Sonora State (Northwest of Mexico) was proposed as the construction site of the north plant, as shown in Figure 1.

Sonora is located within the North West Coastal Plain, which forms a belt 1400 km long, bounded on the east by the Sierra Madre Occidental and on the west by the Gulf of California. It is 250 km wide to the north (Sonora) and 75 km wide to the south (Sinaloa State), with an average elevation of 100 m. The region is mostly flat with a gentle slope towards the sea, interrupted by deeply eroded hills and low mountains or hills surrounded by low lying alluvial plains. From the northern border to the Rio Yaqui, there are large areas of typical desert plains, arreicas and criptorreicas where one can find sand dunes in a half moon.

Figure 1. State of Sonora (source: IMADES)

1. *Climate of the State of Sonora*: The State of Sonora has a dry climate (Figure 2), with an average temperature of 20 °C in the valleys and along the coast, while in the mountain region is 16 °C, with highs of 56 °C and minimum of -10 °C. The northern part of Sonora is characterized by a dry desert climate in the plains near the coast, a temperate rainforest in the mountainous region and the remaining dry steppe. The annual precipitation is 50 to 350 mm in the northwest and 400 to 600 mm in the rest of the state. In the southern desert, the climate is dry and very warm, with a rainfall of 266 mm in the summer.

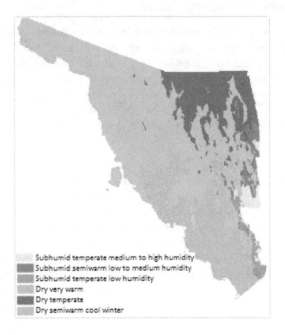

Figure 2. Schematic with different climates in the state of Sonora. (Source: IMADES)

2. *Vegetation of the State of Sonora*: Bushes occupy the largest area of the state (38.07%), dominated by ranching and the removal of wildlife for commercial purposes (mesquite, oregano, chiltepin) and crafts (ironwood, etc.). The areas with no apparent use in extension are next in prominence (17.29%). Pastures are predominantly livestock areas and occupy 13.06%. Forests cover 12.57% of the state and are located in the Sierra Madre Occidental; they are characterized as pine-oak, oak-pine and pine, and although there is infrastructure, the forestry operations remain mostly artisanal. Value-mining areas are well distributed in the state (6.28%) and dominated by gold and copper deposits. Approximately 6.01% is suitable for livestock, and the agricultural areas of the state (4.89%) are mostly irrigated. Intensive livestock (poultry, swine, dairy farms and feedlots) occupies a small area (1.15%), although it is economically important

3. *Soil salinity in the State of Sonora*: Although 90.77% of the territory has no saline problem, approximately 10% is affected by salts at different levels. While 62.3% of the agricultural soils are normal, the rest are either saline-sodic (15.9%), have problems of salinity (12.4%), are strongly saline (3.2%), are strongly saline-sodic (2.3%), have only sodicity problems (1.7%) and are strongly sodic, are moderately saline (0.8%), or are strongly saline (1.4%). Salinity problems primarily exist in the Irrigation Districts of Caborca and the Hermosillo Coast and in the Yaqui and Mayo Valleys. The most important and difficult to eradicate are the saline-sodic soils found in the Yaqui Valley and in the saline delta plain (plains of San Luis Rio, Colorado).

4. *Stationary sources of air pollution*: Using the Information System Rapid Environmental Impact Assessment (SYRIA) simulation model, pig farms, landfills, urban centers, mines, mining and industry emissions were analyzed. In some cases, emission factors were used. The nine municipalities in the State of Sonora comprise nearly 85% of the population and nearly 65% of the productive activities, propagating a proportional burden of pollutants in the atmosphere, which presumably generate 251.2 Mg/year of total particulate matter, 48,037.8 Mg/year of hydrocarbons, and 399.5 Mg/year of carbon oxides in different composition.

5. *Watershed*: The State of Sonora has 12 watersheds, with most domestic consumption taking place in the Sonora River Basin, which passes through the state capital and crosses some of the oldest villages. Next in order of importance are the Yaqui River and Mayo River Basins; they have larger concentrations of people due to an agricultural boom resulting from the construction of hydraulic works. This assertion is reflected in the consumption of water for agricultural activities; water consumption from the Rio Yaqui and Mayo has increased since the construction of the Alvaro Obregon and Adolfo Ruiz Cortines dams. From the point of view of industrial development, these three hydrologic regions also contribute to the increased water consumption and increased demands on the service sector.

2.4.2. Socioeconomic and cultural components

The demand for services and natural resources is determined by the quality of life, which translates to economic growth. To evaluate this demand, we analyzed data on the growth of different economic sectors. The employed population has remained constant over the past three decades. Although the EAP has increased from 25.9% in 1970 to 45% in 1990, the unemployment rate increased from 0.75% in 1980 to 2.5% in 1990. The tertiary sector is the most dynamic in the state, occupying 49%, while 23% are in the primary sector, down from 1970 to 1990. Among the major indigenous groups are the Opata, Yaqui, Papago, Pimas and Seris.

Productive activities (Figure 3) were analyzed based on natural resources, particularly vegetation, as resources show the utilization of the soil. This enables the observation of impacts or consequences of productive activities on the environment.

As mentioned above, the vegetative plane incorporated activities such as aquaculture and was derived from recent satellite imagery. Human settlements and industries were added based on INEGI corrected plans for the 72 largest settlements in the state. Intensive livestock dairies, feedlots, poultry and swine were charted by obtaining the coordinates of each of those registered in the Ministry of Livestock Development, the State Delegation of the Ministry of Agriculture or livestock associations. The mining districts were mapped on the basis of records provided by the Mining Development Division of the Ministry of Economic Development and Productivity.

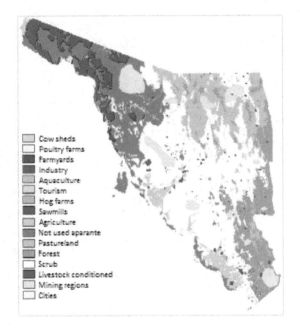

Figure 3. Distribution of land use in the State of Sonora. (Source: IMADES)

2.4.3. Technology and management components

The state of Sonora has a series of dirt roads, paved roads, highways and railroads linking major cities. As a border state, the highway goes straight to the border in Nogales. There is a pipeline that runs through much of the center of the state, which originates in the U.S., passes by Naco, Cananea, and reaches Hermosillo following the route of the highway. This is a great advantage because this is the area that receives the largest amount of solar radiation in the country. Table 2 compares the proposed systems

System Determinants	Traditional System	Hybrid System
Generation Capacity	300 MW combined cycle	50 MW solar thermal (PT) 250 MW combined cycle
Primary energy used	natural gas (imported)	solar energy and natural gas (imported)
Gross efficiency of c.c.	46.79%	46.79%
Net efficiency of c.c.	45.38%	45.38%
DSG system net efficiency with Parabolic trough	-----	23%
Generation time with natural gas	5694 h/year (p.f.= 0.65)	5694 h/year (p.f.= 0.65 for c.c.) 13.2 h/day (85.3%)
Generation time with solar power	-----	2445 h/year* 6.7 h/day (14.7%)
Solar radiation design	-----	2772 kWh/m²
Maximum solar radiation	-----	3122 kWh/m²
Natural gas heat value	9200.14 kcal/m³	9200.14 kcal/m³
Solar concentration area	-----	570 265 m²
Required total area	846 476 m²	1 316 741 m²
Life	30 years	30 years
Domestics inputs	37.4%	53.7%
Imported inputs	62.6%	46.3%
Funding	CFE and private investment	CFE, World Bank, private investment and GEF

* The combined cycle always works and the plant is at full capacity for only 6.7 hours per day.

Where: c.c. --- Combined cycle p.f. --- Plant factor (taken from COPAR) P.T. --- Parabolic trough

Table 2. Comparison of the characteristics of each system.

The plant can be located in any area close to the west of the Hermosillo highway linking the city of Santa Ana south of Nogales. The area around the pipeline from the United States also spans a river, and the amount of solar radiation is the highest in the country. Furthermore, the ground is flat and semiarid.

2.4.4. Identification of critical points in the system

To identify the critical points in the system we ask the following question: What are the environmental factors or processes–technical, social and economic–that individually or in combination may have a crucial effect on the survival of the management system?

a. Environmental aspects: From this point of view, factors that can influence the sustainability of the management system include the following: air pollution from natural gas leaks; the large land area required for the installation of parabolic trough concentrators; large loss of cooling water for weather at certain times of year; low yields from cloudy weather; emissions from burning natural gas; change of land use; erosion; etc.

b. Socioeconomic aspects: These aspects are highly dependent on electricity prices because instability will affect the entire future of a plant using imported natural gas. Though this natural gas pipeline is national, it remains very sensitive to the unit price. International borrowing may be required to finance the construction of any system, and the construction may entail in a high migration of population for the construction of the plant, resulting in an imbalance in the surrounding communities. Combined cycle technology requires a great deal of imported equipment and will be subject to prices quoted in dollars or euros. The cost of labor will be slightly higher compared to the rest of the country, as this is a region near the border. Additionally, the use of an alternate source of energy can create suspicion among investors. Recently, CFE tender-type parabolic trough plants have been deserted both in Agua Prieta and Puerto Libertad (Sonora) due to administrative –not technical– reasons.

2.5. Selection of indicators

To define the indicators used in this evaluation, we must select those that are inclusive, i.e., those that describe rather than analyze processes. The indicators must be easy to measure, easy to obtain and be appropriate for the system under analysis. They must be applicable in a defined range of ecological, socioeconomic and cultural conditions and have a high level of reliability. These indicators should be easy to understand for most readers and be able to measure changes in system characteristics over time in a practical and clear manner. Finally, the measurements must be repeatable over time.

For this paper, we will consider three areas of evaluation: economic, social and technical/environmental, placing the general attributes of sustainability in each of the areas proposed by their own diagnostic criteria.

2.5.1. Economic indicators

To select these indicators, we must first state the diagnostic criteria to follow and then the indicator to use for defining the general attributes of sustainability.

• Productivity: You can assess profitability and efficiency indicators by using Net Present Value, Internal Rate of Return and Cost/Benefit. Other indicators may include the investment cost, turnaround time, etc.

• Stability, Resilience and Reliability: We can assess the diversification of fuel use and risk measurement mechanisms, using indicators for credit, insurance, leverage, percentage of income derived from the use of different primary energies, price of natural gas, etc

- Adaptability: We can evaluate the options of primary energy use and technology options by indicating of number and type of primary energy options and technologies available, the cost at low loads, cost of generation on cloudy days, low demands, etc.

- Equity: The diagnostic criterion is the adaptability of technology and employment trends using indicators such as the cost of investment/production revenue, number of jobs created (temporary and permanent), access to fire insurance, etc.

- Self-Reliance: The indicators measure the level of self-financing, the degree of indebtedness, domestic savings, percentage of self-produced energy use, etc.

2.5.2. Technical and environmental indicators

These indicators give us information about the ability of the proposed systems to be environmentally "productive." Sustainability must sometimes include indicators describing the state of the environment or the processes of prevention and protection of environmental degradation.

- Stability, Resilience and Reliability: Can be used as an indicator of land use patterns, number of species in the area, soil quality and water, soil degradation, disasters, climate change, soil chemical properties, physical soil properties, distribution of natural capital in each region, and so on.

- Self-Reliance: This requires indicators of energy subsidy, energy efficiency and degrees of external dependence.

2.5.3. Social indicators

This type of indicator is very difficult to quantify, especially for a production plant that will supply power to the communities and surrounding cities in Sonora. A much larger study is required to determine the exact number of beneficiaries and the investment schemes to be used for construction and operation. Some of the indicators suggested in the literature (Masera, 1996) are as follows:

- Equity: The distribution of benefits can be used as an indicator of the number and type of benefits by gender, social sector, age, ethnicity, etc. The factors influencing decision-making may include policies, group's resistance, lobbyists and others.

- Stability, Resilience and Reliability: The ability to overcome serious events can affect the survival of the project after conflicts, problems or lack of financing. The processes of learning and training can reference the type and frequency of training, knowledge sharing mechanisms between members, etc.

- Adaptability: Human resource development can be evaluated using indicators such as concepts, methodology and ownership by the community as a capacity for change. We can assess changes in objectives, projects, personnel, and adaptation to changes in the different aspects of production, etc.

• Self-Reliance: Participation is evaluated by the number and frequency of the different phases of the project. We measure the power control that decides on critical aspects of the organization with respect to the type, structure and permanence of the organization.

2.6. Measurement and monitoring indicators

The above indicators will be evaluated quantitatively or qualitatively, depending on the question, because some of them will be justified by arguments or theoretical reasons, partly due to the difficulty in assigning a number to an assessment of non-numeric type. First, we will present the economic indicators, followed by the technical-environmental indicators and then social indicators.

2.6.1. Economic indicators

Regardless of the general attributes of evaluating sustainability indicators, they are calculated individually to reach a conclusion. Table 3 shows the results of these calculations (Geyer, et al., 2004).

Indicator	Conventional System	Hybrid System
Generation [GWh]	1,708	1,708
Investment cost [USD]	$ 135,000,000.00	$ 237,316,931.13
Fuel cost [USD]	$ 68,505,647.32	$ 57,088,039.43
Operation and Maintenance Cost [USD]	$ 8,634,951.00	$ 8,334,592.50
Unit Cost of Generation [USD/kWh]	$ 0.045	$ 0.038
Unit Cost of Investment [USD/kWh]	$ 0.079	$ 0.139
Internal Rate of Recovery	26.90%	19.60%
Net Present Value [USD]	$ 38,943,485.89	$ 33,155,291.46
Annuity equivalent [USD]	$ 7,821,505.70	$ 6,659,196.56
Benefit/Cost	4.11	2.61
Recovery period	4.8 years	3.1 years

Table 3. Economic variables of each model (Source: own data).

The economic and financial analysis necessary to reach these results was performed on a spreadsheet, considering each year of construction, testing, operational development and economic variables.

For economic indicators, we used data from the Costs and Benchmarks for Formulation of Investment Projects in the Electricity Sector - Generation (CFE, 2007) for the combined cycle units, while the thermal data were taken from Hertlein et al. (1990) and Franz Trieb (2009).

2.6.2. Technical and environmental indicators

The location of the proposed plant is 73 km northwest of City of Caborca, off Highway 37 (60 km in a straight line). The vegetation of the area consists of scrub and grassland as non-endemic species. The fauna consists primarily of rodents, reptiles and insects. No crops are grown in the area.

The degradation phenomena studied consisted of erosion, salinity and pollution (soil, water and air). In most cases, estimates were made in the absence of available information by using mathematical models with the aid of GIS and satellite imagery to update information. The erosion in the area lies between 4 and 10 ton/ha. In terms of salinity, the area is within the affected soils. There are no landfills in the vicinity.

The selected region has few clouds for most of the year. However, as the present electricity generation derives from natural gas, the combined cycle plant will operate continuously and the total capacity of the plant will be operational within the CSP.

The energy subsidy should be completely designated as external and not just for the region but also for the nation, as the pipeline that feeds the plant comes from the United States. The energy efficiency is the highest in Mexico. The total conversion efficiency of the solar thermal power plant varies from 21 to 23%, and it can be significantly improved if a Direct Steam Generation solar field is used to deliver steam at 550 °C and 100 bar (Zarza, 2004). The CO_2 emissions from each of the proposed systems are shown in Table 4.

Indicator	Conventional System	Hybrid System
Amount of fuel	386 181 018 m³/year	337 125 816 m³/year
Amount of CO_2 emitted	275 844 ton/year	240 804 ton/year

Table 4. Atmosphere emissions for each model.

2.6.3. Social indicators

The indigenous groups in Sonora who live around the area proposed for the construction of the plant (the Opata, Yaqui, Papago and Pimas) subsist primarily through activities such as the manufacture of handicrafts, animal husbandry and subsistence farming. The construction of the plant would mean a source of temporary employment for them, bringing benefits in both economy and quality of life.

The primary operator will be CFE, although it is very likely that contractors will decide less important aspects that may have a major impact on the region.

In the event of any social conflict in the area, the construction phase will have to stop for security reasons. However, if the pipeline continues to provide natural gas, then the plant will continue producing. There is sufficient security in the area to guard against rebel groups tampering with transmission towers or the pipeline.

Staff at the plant must be local people who receive sufficient training in all aspects of the plant; this can provide a continuity of (very general) knowledge to the community.

The organization of the plant will likely come from CFE because it is the institution that controls and manages the production of electricity. They already have defined organizational schemes in place for the initial production of a new power plant.

The beneficiary communities surrounding the hybrid plant, including the cities of Hermosillo and Santa Ana, are not intended to affect social stability during the construction period (1 to 3 years) and afterwards during operation.

2.7. Evaluation results

The objective of the proposed hybrid system is the generation of electricity for the area by installing the latest technology, meaning that the technology used has the highest possible conversion efficiency of primary energy, that the solid and liquid waste emissions are minimized, and that the system incorporates the additional use of a renewable energy source. According to calculations above, the obtained results show that the amount of generated CO_2 is less than that emitted by the conventional system; hence, the risk for environmental pollution is reduced for future generations. The plant is also adapted to an area with poor socio-ecological circumstances, which will dramatically improve the benefits for future generations.

Production structures for generation, distribution and consumption will provide electrification services and reliable energy necessary for the progress of the region, which facilitates total employment and meaningful work, thereby improving human capabilities for the inhabitants of the region.

With the launch of a hybrid power plant (solar combined cycle), low resource consumption technology is developed that adapts to local socio-ecological circumstances. Because the primary energy sources are low-polluting solar energy and natural gas, there are still significant risks for the present and future. Increasing the electrical infrastructure in the region by consuming imported gas will not preserve this resource for future generations for other uses. It will, however, conserve resources in the zone if outside resources are used.

We must ensure the satisfaction of some of the most basic human needs, such as the provision of high-quality energy. By implementing an innovative system of this type, we promote cultural diversity and pluralism; by using new commercial technology, we can share experiences with other international institutions on the construction of field parabolic trough concentrators and on the development of the plant and its operation.

This should help to reduce the aspects that make it less sustainable to allow the use of more solar energy as renewable primary energy.

In terms of economic indicators such as initial investment, it is slightly more expensive to implement a hybrid plant. However, costs would be absorbed by international organizations and the CFE, whose resources are governmental and therefore contributed by people from across the country. Despite this, the remaining features and sustainability objectives are met; therefore, it can be said that a sustainable system has a very promising future. The advantages outweigh the disadvantages, and thus it is worth implementing a solar power plant for generating electricity under the study conditions. It is a sustainable project from a technical, ecological and social standpoint, but from an economic point of view, it is not entirely sustainable because it requires external resources, which does not comply with the characteristics of self-sufficiency.

As a complement to the results of this analysis, there is an internal report from the Institute of Engineering (Almanza, et al. 1990) that offers a more formal and technical evaluation of the climatic conditions of the proposed area.

3. Conclusions

The amount of CO_2 emission from the hybrid system is less than the conventional system. Production structures are generated, and the plant develops hybrid technology with a low consumption of resources, which are adapted to local socio-ecological circumstances.

Since there is a wide acceptance in all social sectors of the concept of sustainability, the proposal made in this paper aims to ensure a sustainable supply of high quality electricity to a region where one of its main natural resource is the Sun spite that, today, still we can not objectively define that term and therefore implement it.

It is very important to make clear that sustainability goes beyond ensuring the environmental integrity of a site and the standard of living of a population, should address the concept of "quality of life" and a form of collective life.

Sustainability is now in a further process of discourse, and efforts to measure it should become a state priority. Institutional initiatives and debates about the measurement of sustainability in general show resistance in committing to this concept. Therefore, there is no common shared understanding of what has been measured.

Sustainability indicators are often an amalgam of economic, social and environmental indicators, but recently, they are showing signs of maturity with better measures of sustainability. These indicators, however, are limited and may reflect unsustainable measures. Their primary value is to indicate the direction of change rather than a swing state.

The indicators are just the initial map and not what could be called the territory. The difficult task of achieving sustainability is another issue.

In consulting references, it follows that the most successful initiatives in measuring sustainability are those initiated and controlled by autonomous public groups (e.g., Sustainable Seattle 1998), where the process is more important than the indicators.

The greater the public involvement in the execution of a community role (for example, consensus conferences, citizen juries, etc.), the more likely we are to achieve sustainability.

We need to address the fundamental existential questions and find meaning in life if we are to achieve sustainability.

The emergy evaluation assigns a value to products and services through their conversion into an equivalent form of energy: solar energy (Odum, 1983, 1996).

Solar energy is used as the common denominator through which different types of resources, whether energy or material, can be measured and compared with others.

Author details

Rafael Almanza and Iván Martínez

Universidad Nacional Autónoma de México / Universidad Autónoma del Estado de México, México

References

[1] Almanza, R, Valdés, A, Lugo, R, Zamora, J M, Sandoval, C y Estrada Cajigal, V. (1990). *Estudio del comportamiento de sistemas solares térmicos para generar electricidad en 5 comunidades de Sonora o Baja California*. Proyecto 0102 Instituto de Ingeniería, UNAM. Patrocinado por CFE. (abril 1990) pp. (1-254)

[2] Boswell, M.R. (1995). Establishing indicators of sustainable development. *Proceedings of the Annual Conference of the Association of Collegiate Schools of Planning*. Detroit, Michigan, USA. October 1995.

[3] Brundtland, G. (Chairman). (March, 1987). Report of the World Commission on Environment and Development: *Our Common Future*. Available in: http://www.un-documents.net/wced-ocf.htm

[4] Cobb, C., Halstead, T. and Rowe, J. (1995). If the GDP is up, why is America down?, In: *Atlantic Monthly digital edition*, Available from: http://www.theatlantic.com/past / politics/ecbig/gdp.htm

[5] Comisión Federal de Electricidad (CFE). (2007). Costos y Parámetros de Referencia para la Formulación de Proyectos de Inversión en el Sector Eléctrico – Generación. (April 2007), México.

[6] Daly, H. and Cobb, J. (1994) *For the Common Good: Redirecting the Economy toward Community, the Environment, and a Sustainable Future*. (2nd edition). Beacon Press. ISBN 9780807047057. USA.

[7] Demirbas, A. (2000). Biomass resources for energy and chemical industry. In: *Energy Education, Science and Technology*. Vol. 5, pp. (21–45)

[8]   Flamant, G., Kurtcuoglu, V., Murray, J., Steinfeld, A. (2006). Purification of metallurgical grade silicon by a solar process. *Solar Energy Materials & Solar Cells*. Vol. 90, No. 14, (September 2006), pp. (2099–2106), ISSN 0927-0248.

[9]   Gligo, N. (1995), Situación y perspectivas ambientales en AméricaLatina y el Caribe, *Revista de la CEPAL*, No 55 (LC/G.1858-P), Santiago de Chile.

[10] Griffin, J. (Ed.) (2007). *World oil outlook 2007*. Organization of t Petroleum Exporting Countries, ISBN: 9783200009653, Vienna, Austria.

[11] Hang, Q., Jun, Z., Xiao, Y., Junkui, C. (2008). Prospect of concentrating solar power in China the sustainable future. *Renewable & Sustainable Energy Reviews*. Vol. 12, No. 9, (December 2008), pp. (2505–2514), ISSN: 1364-0321.

[12] Hart, M. (1995). Guide to Sustainable Community Indicators. Ipswich, Maine. QLF/ Atlantic Center for the Environment.

[13] Henderson, H. (1995). The Indicators Crisis, In: *Paradigms in Progress: Life Beyond Economics*, pp. (147-192), Berrett-Koehler Publishers, ISBN 978-1881052746, USA.

[14] Henderzon, H. (1991). Paradigs in progress: Life beyond economics. (2nd Ed.). Knowledge Systems. ISBN 9780941705219. USA.

[15] Hertlein, Klaiss, and Nitsch. (1990). *Cost analysis of solar power plants*. American Solar Energy Society. USA.

[16] Holden, M. (2006). Sustainable Seattle 1998: The Case of the Prototype Sustainability Indicators Project. In: *Community Quality-of-Life Indicators Social Indicators Research Series*. (2006). Vol. 28, pp. (177-201), DOI: 10.1007/978-1-4020-4625-4_7

[17] Liverman, D, Hanson, M E, Brown, B J, y Meridith, R W. (1988). Global sustainability: towards measurement. In: *Journal of Environmental Management*. Vol. 12, No. 2, (1988), pp. 133-143, DOI: 10.1007/BF01873382

[18] Masera, O.; Astier, M. (1996). *Metodología para la Evaluación de Sistemas de Manejo Incorporando Indicadores de Sustentabilidad*. Grupo Interdisciplinario de Tecnología Rural Aplicada, México.

[19] Morse, F. (2008). Concentrating Solar Power (CSP) as an option to replace coal. In: Energy & Climate Mini-Workshop, Washington, DC. 14/05/2012. Available from: http://ossfoundation.us/projects/ energy/2008

[20] Odum, H.T. (1988). Self-Organization, Transformity, and Information. *Science*, Vol. 242, No. 4882, (November 1988), pp. (1132-1139), ISSN: 0036-8075 (print), 1095-9203 (online).

[21] Organization of the Petroleum Exporting Countries (OPEC). (2007). *World oil outlook 2007*. Technical Report OPEC. ISBN: 9783200009653, Vienna, Austria.

[22] Patlitzianas, K., Kagiannas, A., Askounis, D., Psarras, J. (2005). The policy perspective for RES development in the new member states of the EU. *Renewable Energy*. Vol. 30, No. 4, (April 2005), pp. (477–492), ISSN: 0960-1481.

[23] Rees, W., Wackernagel, M., and Testemale, P. (1998). *Our Ecological Footprint: Reducing Human Impact on the Earth*. New Society Publishers, ISBN 9780865713123, Canada.

[24] Sachs, W. (1996). What kind of sustainability?, *Resurgence*, No 180, (1996), pp. (20-22).

[25] Scienceman, D.M. and El-Youssef, B.M. (1993). The System of Emergy Units. *Proceedings of the XXXVII annual meeting of the International Society for the Systems Sciences*. New South Wales, Australia, July 1993.

[26] Trieb, F., Schillings, C., O'Sullivan, M., Pregger, T. and Hoyer, C. (2009). Global Potenctial of Concentrating Solar Power. *Proceeding of the 15th Solar PACES International Symposium on Concentrated Solar Power and Chemical Energy Technologies*. Berlin, Germany, September 2009.

[27] Trzyna, T. (1995). A Sustainable World; Definig and measuring Sustainable Development. IUCN. Earthscan.

[28] Trzyna, T. (Ed). (1996). *A Sustainable World: Defining and Measuring Sustainability*. International Union for Conservation of Nature, ISBN: 978-1880028025, Switzerland.

[29] World Commission on Environment and Development. (1987). *Our Common Future*. Oxford University Press. ISBN 9780192820808. Great Britain.

[30] Yanarella, E.J. and Levine, R.S. (1992). Does sustainable development lead to sustainability?, *Futures*. Vol. 24, No. 8, (October 1992), pp. (759-774), ISSN: 0016-3287.

[31] Yüksel, I. (2008). Global warming and renewable energy sources for sustainable development in Turkey. *Renewable Energy*. Vol. 33, No. 4, (April 2008), pp. (802–812), ISSN: 0960-1481.

[32] Zarza, E., Valenzuela, L., León, J., Hennecke, K., Eck, M., Weywers, H., Eickhoff, M. (2004) Direct Steam Generation in parabolic troughs, final results and of the DISS project. *Proceeding of the 12th Solar PACES International Symposium on Concentrated Solar Power and Chemical Energy Technologies*. Oaxaca, México, October 2004

Chapter 3

Proof of the Energetic Efficiency of Fresh Air, Solar Draught Power Plants

Radu D. Rugescu

Additional information is available at the end of the chapter

1. Introduction

The thermal draft principle is currently used in exhaust chimneys to enhance combustion in domestic or industrial heating installations. An introductory level theory of gravity draught in stacks was issued by the old German research institute for heating and ventilation (Hermann-Rietschel-Institut) in Charlottenburg, in a widely translated reference book (Raiss 1970). Technological and practical aspects of air draught management are clearly exposed in this works, but a wide-predicting theory still lacks. As early as in 1931 a surprisingly advanced proposal to use thermal draught as a propelling system to generate electricity from solar energy was forwarded by another German researcher (Günter 1931). Major advancements in convective flows prediction during the last decades of the 20th century were accompanied by a series of publications and we cite first the basic book due to a work from Darmstadt (Unger 1988). The related topic of convective *heat transfer*, often involved in thermal draught, was also intensively studied and the advanced results published (Jaluria 1980; Bejan 1984). With these records the slippery analytical theory of natural gravity draught was set well under control. Thermal energy from direct solar heating is regularly transformed into electricity by means of steam turbines or Stirling closed-loop engines, both with low or limited reliability and efficiency (Schiel et al. 1994, Mancini 1998, Schleich 2005, Gannon & Von Backström 2003, Rugescu 2005). Steam turbines are driven through highly vaporised water into tanks heated on top of supporting towers, where solar light is concentrated trough heliostat mirror arrays. High maintenance costs, the low reliability and large area occupied by the facility had dropped the interest into such renewable energy power plants. The alternative to moderately warm the fresh air into a large green house and draught it into a tower, checked only once, gave also a very low energetic efficiency, due to the modest heating along the green house. This existing experience has fed up a visible reluctance towards the solar tower power plants (Haaf 1984).

However, a simple and efficient solution exists which is here demonstrated by means of energy conservation. This method provides a superior energetic efficiency with moderate costs and a high reliability through simplicity. It consists of optimally heating the fresh-air by means of a mirror array concentrator and an efficient solar receiver, and accelerating it further in the tall towers through gravity draught (Fig. 1, Rugescu 2005).

Figure 1. Project of the ADDA solar array gravity draught accelerator.

This genuine combination has already a history of theoretical study (Rugescu 2005) and an incipient experimental history too (Rugescu et al. 2005). First designed for air acceleration without any moving parts or drivers with application to infra-turbulence aerodynamics and aeroacoustics, the project was further extended for green energy applications along a series of published studies (Rugescu et al. 2006, Rugescu 2008, Rugescu et al. 2008, Rugescu et al. 2009, Rugescu et al. 2010, Cirligeanu et al. 2010, Rugescu et al. 2011a, Rugescu et al. 2011b, Rugescu 2012, Rugescu et al. 2012a, Rugescu et al. 2012b). The demonstration of the high draught tower energetic efficiency provided below is expected to convince the skeptics and to bolster again the direct solar energy exploitation in tall tower power plants (Rugescu et al. 2012 b).

2. Gravity-draught accelerator modeling

A schematic diagram of a generic draught tower is drawn in Fig. 2. The fresh air in its ascending motion up the tower, due to the gravity draught, is first absorbed, from the immobile atmosphere ($w_0=0$), through the symmetrically positioned air intakes at the level designated as station "0", close to the ground (Fig. 2). It turns upright along the curved intake and accelerates afterwards to the velocity w_1 through the laminator. It then enters the solar heater, or solar receiver, at station "1" into the stack. Due to warming and dilatation into that receiver by absorption of the thermal flux $\dot{q}$ it accelerates further to velocity w_2 at receiver exit "2", from where after the heat transfer to the walls is small and is supposedly neglected and the light air is draught upwards with almost constant velocity up to the upper exit of the tower "3", under the influence of the differential gravity effect of almost constant intensity g between the inner and outer zone of the atmosphere. The tower secures an almost one-directional flow and consequently the problem will be treated here as one-dimensional.

The ideal gas behavior under the influence of a gravity field of intensity g, flowing upward with the local velocity w into a vertical duct of cross area A and subjected to a side wall heating by a thermal flux $\dot{q}$ is fully described by the 3-D conservation laws of mass, impulse, energy, by the equation of state and by the physical properties of the gas, the air in particular.

The air flow of the material, infinitesimal control volume $dV \equiv A(x)\,dx$ into the vertical pipe of variable cross area A and subjected to side heating by a thermal flux $\dot{q}(t, x)$ is described by the conservation laws of mass, impulse and energy successively:

$$\frac{dM_V}{dt} \equiv \frac{d}{dt} \int_{dV} \rho dV = \frac{\partial}{\partial t} \int_{dV} \rho dV - \oint_{\partial dV} w \cdot n \, \rho dS = 0 \tag{1}$$

$$\frac{dH_V}{dt} \equiv \frac{d}{dt} \int_{dV} \rho w dV = \frac{\partial}{\partial t} \int_{dV} \rho w dV - \oint_{\partial dV} w\, w \cdot n \, \rho dS = \oint_{\partial dV} \tau \cdot n \, dS + \int_{dV} \rho g \, dV \tag{2}$$

$$\frac{dE_V}{dt} \equiv = \frac{\partial}{\partial t} \int_{dV} \left(e + \frac{w^2}{2}\right) \rho dV - \oint_{\partial dV} \left(e + \frac{w^2}{2}\right) w_n \, \rho dS = \oint_{\partial dV} \tau \cdot n \cdot w \, dS + \oint_{\partial dV} \dot{q} \, dS + \int_{dV} g \cdot w \, \rho \, dV \tag{3}$$

where e and k are the intensive inner energy and kinetic energy of the gas, respectively. The stress tensor τ acts on the walls only, meaning the boundary of the control volume.

The computational solution of the stack flow further depends on the initial and limit conditions that must fit the physical process of thermal draught (Bejan 1984) and may be managed in simple thermodynamic terms. In its general form, the dynamic equilibrium of the stack flow was first debated in a dedicated book (Unger 1988), with emphasize on the static pressure equilibrium within and outside the stack at the openings, the key of the entire stack problem. The one-dimensional steady flow assumption with negligible friction was accounted and we add the proofs that this approach is consistent with the problem. In that regard we analyze in a new way the flow with friction losses, estimate their magnitude and add a different accounting for compressibility at entrance. Our point of view faintly modifies the foregoing results regarding the compressibility of the air during inlet and exit acceleration, still consists of a necessary improvement.

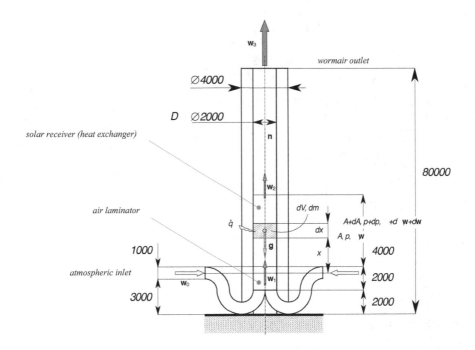

Figure 2. Control volume into a generic stack.

The aerostatic influence of the gravitation is then given by the pressure gradient equation inside (density ϱ) and outside (density ϱ_0) the tower,

$$\frac{dp}{dz} = -g\,\rho, \qquad \frac{dp}{dz} = -g\,\rho_0 \tag{4}$$

The right hand term in these equations is nothing but the slope to the left of the vertical in each pressure diagram from Fig. 3.

This means that the inner pressure in the stack (left, doted line) is decreasing less steeply and remains closer to the vertical than the outer pressure of the atmosphere. The dynamic equilibrium is established when, following a series of transforms, the stagnation pressures inside and outside become equal (Fig. 3). While the air outside the stack preserves immobile and due to the effect of gravitation its pressure decreases with altitude from $p_{ou}(0) \equiv p_0$ at the stack's pad to $p_{ou}(\ell)$- at the tip of the stack "4", the inner air is flowing and consequently its pressure p_{in} varies not only by gravitation but also due to acceleration and braking along the 0-1-2-3-4 cycle.

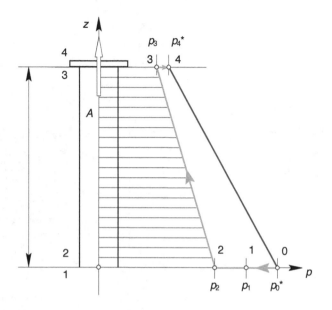

Figure 3. Dynamics of the gravitation draught.

Under the assumption of a slender tower with constant cross area A, meaning a unidirectional flow under an established, steady-state condition with friction under laminar behavior or developed turbulence, the conservation laws for a finite control volume from stage "1" to station "z" are further developing into the conservation of mass,

$$\rho w = const. \tag{5}$$

and energy for the compressible flow, with the assumption $\rho_1 \approx \rho_0$,

$$\frac{p_0}{\rho_0} = \frac{p_1}{\rho_1} + \frac{\kappa - 1}{2\kappa} w_1^2 \leftrightarrow p_0 = p_1 + \frac{k}{2} \frac{\dot{m}^2}{\rho_0 A^2} \tag{6}$$

for the entrance into the stack. In other words the air acceleration takes place at tower inlet between 0-1 as governed by the energy compressible equation with constant density ρ_0 along,

$$p_1 = p_0 - \frac{\Gamma}{2} \cdot \frac{\dot{m}^2}{\rho_0 A^2} \tag{7}$$

where A is the cross area of the inner channel, $\dot{m}$ the mass flow rate, constant through the entire stack (steady-state assumption) and the thermal constant Γ with the value for the cold air

$$\Gamma \equiv \frac{\kappa - 1}{\kappa} = 0.28826 \tag{8}$$

The air is warmed in the heat exchanger/solar receiver between the sections 1-2 with the heat q per kg with dilatation and acceleration of the airflow, accompanied by the "*dilatation drag*" pressure loss. Considering again $A=const$ for the cross-area of the heating zone too, the continuity condition shows that the variation of the speed is simply given by

$$w_2 = w_1 / \beta \tag{9}$$

The impulse equation gives now the value of the pressure loss due to air dilatation,

$$p_2 + \frac{\dot{m}^2}{\rho_2 A^2} = p_1 + \frac{\dot{m}^2}{\rho_0 A^2} - \Delta p_R \tag{10}$$

where a possible pressure loss into the heat exchanger Δp_f due to friction is considered. Once the dilatation drag is thus perfectly identified, the total pressure loss Δp_Σ from pad's outside up to the exit from the heat exchanger results as the sum of the inlet acceleration loss (7) and the dilatation loss (10),

$$p_2 = p_0 - \frac{\dot{m}^2}{2\rho_0 A^2} + \frac{\dot{m}^2}{\rho_0 A^2} - \frac{\dot{m}^2}{\rho_2 A^2} - \Delta p_R \equiv p_0 - \Delta p_\Sigma, \tag{11}$$

equivalent to

$$p_2 = p_0 - \frac{\dot{m}^2}{\rho_0 A^2} \cdot \frac{r(2-\Gamma)+\Gamma}{2(1-r)} - \Delta p_R \tag{12}$$

The gravitational effect (4) continues to decrease the value of the inner pressure up to the exit rim of the stack, where the inner pressure becomes

$$p_3 \equiv p_2 - g\rho_2 \ell = p_0 - \frac{\dot{m}^2}{\rho_0 A^2} \cdot \frac{r(2-\Gamma)+\Gamma}{2(1-r)} - \Delta p_R - g\rho_2 \ell \tag{13}$$

Either the impulse equation in the form

$$\frac{\dot{m}}{A}(w_2 - w_1) = p_1 - p_{in}(z) - \Delta p_f - g\rho_2 z, \tag{14}$$

or the energy equation in the form

$$\frac{\kappa}{\kappa-1}\left(\frac{p_{in}}{\rho}-\frac{p_1}{\rho_1}\right)+\frac{w^2-w_1^2}{2}=\left(\frac{\dot{q}}{\dot{m}}P_{in}-g\right)(z-z_1) \tag{15}$$

appears for the receiver, heated zone, and

$$p_2-p_{in}(z)=\frac{\kappa-1}{\kappa}g\rho_2(z-z_2), \ p_2-p_3=\frac{\kappa-1}{\kappa}g\rho_2(z_3-z_2) \tag{16}$$

for the free ascending flow above the receiver, with P_{in} for the perimeter length of the inner channel walls.

At the upper exit from the stack the gas is diluting and braking into the still atmosphere, thus the compressible Bernoulli equation applies,

$$p_3+\frac{\kappa-1}{\kappa}\frac{\rho_2}{2}w_3^2=p_{4'} \qquad p_3+\frac{\kappa-1}{\kappa}\frac{\rho_2^2w_3^2A^2}{2\rho_2A^2}=p_{4'} \tag{17}$$

when constant density during this process is assumed again. The pressure variation at stack's exit is very small and this ends in the fact that other simplifying hypotheses do not give results consistent with the physical phenomena.

Modifying eq. (14) the inner static pressure at stage z with friction is immediately delivered into the following expression

$$p_{in}(z)=p_{ou}(z)-\frac{1+r}{2(1-r)}\cdot\frac{\dot{m}}{\rho_0A^2}-\Delta\, p_f(z)+g\,\Delta\rho\, z \tag{18}$$

where the relative heating of the air is expressed in terms of densities,

$$r\equiv\frac{\rho_0-\rho_2}{\rho_0}=1-\frac{\rho_2}{\rho_0}\equiv1-\beta \tag{19}$$

with a given control value for

$$\beta = \frac{p_2}{p_0} < 1 \tag{20}$$

Using eq. (17) the static pressure of the exhausted air becomes

$$p_4(\ell) = p_{ou}(\ell) - \left[\frac{1+r-\Gamma}{1-r} \cdot \frac{\dot{m}^2}{2 p_0 A^2} + \Delta p_f(\ell) - g \Delta \varrho \ell \right] \tag{21}$$

which is used in the equilibrium condition as follows.

The values of the pressures and velocities into the main sections result from the equilibrium condition of the pressures above the upper exit, where the inner $p_4(\ell)$ and the outer $p_4^* = p_{ou}(\ell)$ values should be equal. That means the square bracket in (21) is set to zero.

In this way (Unger 1988, Rugescu 2005, Rugescu et al. 2005), the mass flow rate through the stack mainly depends on the relative heating of the air, expressed in terms of densities, and results when the pressure difference between the interior and the exterior of the tower exit recovers by dynamic braking of the air (Fig. 3).

$$\frac{\Delta p(\ell)}{g \varrho_0 \ell} = \frac{1+r-\Gamma}{1-r} \cdot \frac{\dot{m}^2}{2g \ell \varrho_0^2 A^2} + \frac{\Delta p_f(\ell)}{g \varrho_0 \ell} - r = 0 \tag{22}$$

For negligible friction losses ($\Delta p_f(\ell) = 0$) the equilibrium mass flow rate becomes

$$R^2 \equiv \frac{\dot{m}^2}{2g \ell \, p_0^2 A^2} = \frac{r(1-r)}{1+r-\Gamma} \tag{23}$$

slightly higher than the predicted value of the previous models (Unger, 1988).

When the friction losses are considered, the actual value for the quadratic mass flow rate results from the second degree equation (22-12) which gets the form,

$$a \left(\frac{\dot{m}}{\dot{m}_\ell} \right)^2 + b \frac{\dot{m}}{\dot{m}_\ell} - r = 0 \tag{24}$$

where at the nominator a reference free-fall mass flow rate appears, $\dot{m}_\ell = w_\ell p_0 A$,

based on the Torricelli free-fall velocity

$$w_\ell^2 = 2g \, \ell, \tag{25}$$

with the constants

$$a = \frac{r/R}{w_\ell^2 \rho_0^2 A^2}, \quad b = \frac{32\, v_0}{AD^2 g \rho_0} \left(\frac{T_w}{T_c}\right)^{1.7} \tag{26}$$

For an example slender, tall stack with the inner channel of elongation $\ell/D = 70/2$ the resulting contribution of friction is really small,

$b/a = 2\%$,

meaning that the difference from the frictionless flow is actually smaller than 0.5 ‰. Consequently the non-friction result in (23-13) should be considered as accurate. Its quadratic form shows the known fact that the heating of the inner air presents an optimal value and there exist an upper limit of the heating where the flow in the stack ceases.

Formula (23-13) shows that the non-dimensional quadratic mass flow rate R^2 is in fact simply the squared ratio of the exhibited stack entrance speed w_1 over the free-fall speed w_ℓ, due to the constant cross area of the stack,

$$R^2(r) \equiv \left(\frac{w_1}{w_\ell}\right)^2 = \frac{r(1-r)}{1+r-\Gamma}, \tag{27}$$

and is given by

$$R^2(r) \equiv \frac{\dot{m}^2}{w_\ell^2 \rho_0^2 A^2} = \frac{r(1-r)}{1+r-\Gamma} \tag{28}$$

The entrance speed exhibits a maximum at the theoretically optimal heating r_{opt},

$$dR^2/dr = 0, \quad r_{opt}^2 + 2(1-\Gamma)\, r_{opt} - (1-\Gamma) = 0, \tag{29}$$

namely

$$r_{opt} = -(1-\Gamma) + \sqrt{(1-\Gamma)(2-\Gamma)} \tag{30}$$

The optimal heating for the standard air appears at a relative density reduction

$$r_{opt} \equiv (\rho_0 - \rho)/\rho_0 = 0.392033 \rightarrow R^2(r_{opt}) = R^2_{max}, \tag{31}$$

meaning an equal increase of the absolute temperature of $(1+r)$ times, when the normal air temperature should be raised with around 120°C above 27°C to achieve a maximal discharge. Due to Archimedes' effect (Unger, 1988), these values are an optimal response to the craft balance between the drag of the inflated hot air and its buoyant force.

A slightly improved model is delivered when the following conditions at the upper exit are introduced, starting from equation (19). The constant density assumption along the upper stack $\rho_2 = \rho_3 = \rho_4$ was used. Recovery of the static air pressure, previously considered through a compressible process governed by the Bernoulli equation (Rugescu 2005)

$$p_4{}^* = p_3 + \Gamma \frac{\dot{m}^2}{2\rho_2 A^2} \tag{32}$$

is here replaced with the condition (Unger 1988) of an isobaric exit $p_4{}^* = p_3$ which, considered into (19) for replacing p_3, ends in the equilibrium equation

$$p_4{}^* = p_0 - \frac{\dot{m}^2}{\rho_0 A^2} \cdot \frac{r\,(2-\Gamma)+\Gamma}{2(1-r)} - \Delta p_R - g\rho_2 \ell \tag{33}$$

This means that the dynamic equilibrium is re-established when the stagnation pressure from inside the tower equals the one from outside, at the exit level,

$$p_4{}^* \equiv p_{in}(\ell) = p_{ou}(\ell) \equiv p_0(0) - g\rho_0 \ell \tag{34}$$

This equation is the end element that allows determining the equilibrium value of the air mass flow rate passing through the stack. Equaling (20) and (21),

$$p_0 - g\rho_0 \ell \equiv p_0 - \frac{\dot{m}^2}{\rho_0 A^2} \cdot \frac{r\,(2-\Gamma)+\Gamma}{2(1-r)} - \Delta p_R - g\rho_2 \ell \tag{35}$$

Reducing by the quotient $g\varrho_0 \ell$ the equilibrium equation appears in the form

$$\frac{\Delta p(\ell)}{g\,\varrho_0 \ell} = \frac{1+r-\Gamma}{1-r} \cdot \frac{\dot{m}^2}{2g\,\ell\,\varrho_0{}^2 A^2} + \frac{\Delta p_R(\ell)}{g\,\varrho_0 \ell} - r = 0 \tag{36}$$

Depending on the construction of the heat exchanger the drag largely varies. For simple, tubular channels the pressure loss due to frictions stands negligible (Rugescu et al. 2005a, Rugescu 2005, Rugescu et al 2005b) and the reduced mass flow rate (RMF) results from the simple equation

$$R^2 \equiv \frac{\dot{m}^2}{2g\,\ell\,\rho_0{}^2 A^2} = \frac{r\cdot(1-r)}{r\,(2-\Gamma)+\Gamma} \tag{37}$$

It gives an alternative to the previous solution of Unger (Unger 1988)

$$R^2 = \frac{r(1-r)}{1+r},$$ (38)

or to the one from above (Rugescu et al. 2005a)

$$R^2 = \frac{r(1-r)}{1+r-\Gamma},$$ (39)

and gives optimistic values in the region of smaller values of heating (Fig. 4).

The behavior of the chimney flow for various heating intensities of the airflow, in the limit case of equal far stagnation pressures (FSP) and for the three different models described is reproduced in Fig. 4, where the limiting, linear cases of the dynamic equilibrium are drawn through straight, tangent lines. These are in fact the derivatives of the mass flow rate in respect to r for the two limiting cases of heating.

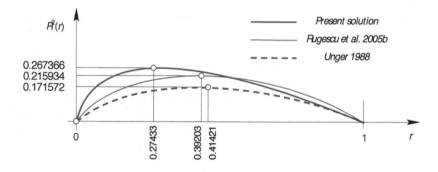

Figure 4. Stack discharge R^2 versus the air heating intensity r.

Differences between the present solution and the previous ones, as given in the above diagram, are non-negligible and show the sensible effect of the variation in modeling of the compressibility behavior at entrance and exit of the stack. This is explained by the tinny variations in pressure and density during the very small acceleration of the air at tower inlet that makes the flow highly sensible to pressure perturbations, either natural or numerical. The same applies for the tower exit. For this reason the previous solution was obtained by completely neglecting the air compressibility at tower upper exit, where the static pressure was taken into consideration instead of the dynamic one.

Numerical simulations of the ducted airflow and the experimental measurements on a scale model support of the present model. The conclusion of this very simplified but efficient modeling of the self-sustained gravity draught, with no energy extraction, is that the heating of the air must be limited to between 0.3÷0.5 in terms of the relative density reduction through heating, or to between 90÷150ºC in terms of air temperature after heating, because under the accepted assumptions the product ϱT preserves almost constant. The optimal heating is thus surprisingly small. The maximum of function in Fig. 4 is flat and the minimal heating limit of 100ºC could be taken as sufficient for the best gravity draught acceleration. Recollection must be made that for the Manzanares green-house power station the air temperature increment was of 20ºC at maximal insolation only (Haaf 1984), fact that explains the failure of this project in demonstrating the ability of solar towers to produce electricity.

The accelerating potential and the expense of heat to perform this acceleration at optimal conditions result from equations (37)÷(39). In a practical manner, the velocity c_2 results in regard to the free-fall velocity (Torricelli) c_ℓ. Its upper margin is given by (40) through (37), while the lower margin by (41) through (38),

$$c_{2H} = \sqrt{\frac{r \cdot 2g\ell}{(1-r)[r\,(2-\Gamma)+\Gamma]}},\qquad(40)$$

$$c_{2L} = \sqrt{\frac{r \cdot 2g\ell}{1-r^2}}\qquad(41)$$

In fact these formulae render identical results for the optimal values for r (Table 1). For a contraction aria ratio of 10 the maximal airflow velocities in the test chamber c_e of the aeroacoustic tunnel versus the tower height are given in Table 1.

ℓ	c_ℓ	c_1	c_2	c_e
m	m/s	m/s	m/s	m/s
7	11.72	4.85	8.28	82.8
14	16.57	6.86	11.72	117.2
30	24.26	10.05	17.15	171.5
70	37.05	15.35	26.20	262.0
140	52.40	21.71	37.05	370.5

Table 1. Draught vs. tower height for a contraction ratio 10.

The value of c_e was computed according to the simple, incompressible assumption, which renders a minimal estimate for the air velocity in the contracted entrance area. Compressibility whatsoever will increase the actual velocity in the test area, while drag losses, especially those in the heat exchanger, will decrease that speed.

3. Experimental results

With the existing small-scale test rig built by the team of University "Politehnica" of Bucharest, the tests that have been conducted led to the values for air velocity in the tube as given in the diagram below. The average values, measured at a distance of 1.7 m from the entrance area of the tube, were registered as 2.115 m/s air speed with the contracted area effect (simulation of a turbine) and of 6.216 m/s without turbine simulation. Air temperature at the exit section was recorded to be of 195°C and 123°C, respectively (Tache et al. 2006).

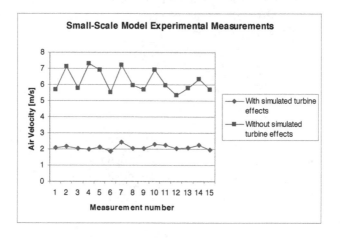

Figure 5. Experimental measurements on the small-scale model

The turbine simulation and the image of the inner electrical heater, simulating the solar receiver, are shown in figures below.

Figure 6. Turbine simulator

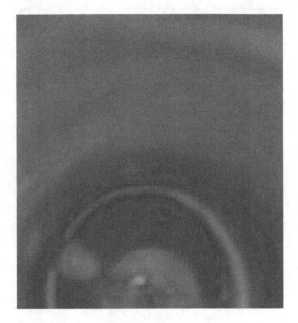

Figure 7. The air heater.

Figure 8. Small-scale model of the draught tower driver(overall view, ¼ contraction area, hot resistors, exit temperature)

The experimental values recorded during the measurement session and the ones obtained from numerical simulations are listed in Table 2.

No.	Measured Air Velocity [m/s]			Simulated Air Velocity [m/s]		
	With contraction	Without contraction	Speed Ratio	With contraction	Without contraction	Speed ratio
1	2.101	5.703	2.714			
2	2.190	7.110	3.247			
3	2.051	5.767	2.811	$V_{min} = 2.19$	$V_{min} = 5.90$	
4	1.996	7.310	3.662			
5	2.127	6.920	3.253			
6	1.867	5.521	2.957			
7	2.414	7.208	2.986			
8	2.027	5.966	2.943			
9	2.051	5.703	2.780			
10	2.307	6.920	3.000			
11	2.276	5.966	2.621	$V_{max} = 3.29$	$V_{max} = 7.07$	
12	2.027	5.351	2.639			
13	2.076	5.767	2.778			
14	2.276	6.329	2.780			
15	1.937	5.703	2.945			
Mean values	2.115	6.216	2.941	2.740	6.485	2.260

Table 2. Experimental and simulated air velocity values

The differences between these values are small, with greater values (~29.55%) when accounting for the turbine effects and much smaller values (~4.33%) in the other case.

4. Design example

As already stated, the optimal air heating for a good draught effect (Fig. 4) stays between 50÷100°C and the computational problem is the following. Given the solar radiance flux, the reflectivity properties of the mirrors and the albedo of the tower walls, find the required area ratio of the solar reflector to the tower cross area that assures the imposed air heating. Considering the optional heating for a good mass flow-rate, formula (30) shows that, near the extreme pick, the discharge rate little depends on the heating intensity r. It was shown in (31) that the optimal rarefaction is placed around $r=0.4$, when the maximal discharge rate of $R^2=0.216$ manifests. Even at a moderate rarefaction of $r=0.14$ only, meaning a 50°C temperature rise above 27°C,

$$\frac{\rho}{\rho_0} \cong \frac{T_0}{T} = \frac{300}{350} \equiv 0.8571, \quad r \equiv 1 - \frac{\rho}{\rho_0} = 1 - 0.8571 \equiv 0.142857, \tag{42}$$

the discharge of the stack exhibits a good value of 2/3 of the maximal one,

$$R^2(r) \equiv \frac{r(1-r)}{1+r-\Gamma} = \frac{0.142857 \cdot 0.8571}{1 + 0.142857 - 0.288256} \equiv 0.1433 \tag{43}$$

At half of the optimal heating, that means at 100ºC, the discharge is comfortably up to 90% of the maximal one, or

$$R^2(r) \equiv \frac{r(1-r)}{1+r-\Gamma} = \frac{0.25 \cdot 0.75}{1.25 - 0.288256} \equiv 0.1950 \tag{44}$$

Under these circumstances it is fairly reasonable to accept for the further computation a moderate rarefaction of r=0.14 or 50ºC heating. With this value and the configuration in Fig. 2, meaning a 2-m internal diameter and again a tower height of 70 meters, the entrance velocity of the air becomes

$$w_1 = \sqrt{2g\ell \cdot R^2} = \sqrt{1372.931 \cdot 0.1433} = 14.03 \ m/s, \tag{45}$$

where the density of the air is still the normal one ρ_0=1.225kg/m^3. Then the mass flow rate equals the value of

$$\dot{m} \equiv \rho_0 w_1 A = 1,225 \cdot 14.03 \cdot 3.1415926 \equiv 54,0 \ kg/s \tag{46}$$

Considering now a rough constant pressure specific heat of the air of

$$c_p = 1005 \ \frac{J}{kg \cdot K},$$

the power consumed with the heating of the air raises to

$$Q_1 \equiv \dot{m} \cdot c_p \cdot \Delta T = 54.0 \cdot 1005 \cdot 50 \equiv 2712498.3 \ W \tag{47}$$

Under a global heating efficiency of 80% the required total solar irradiation is

$$Q \equiv Q_1/\eta = 2.7124983/0.8 \cong 3.39 \ MW \tag{48}$$

The lunar-averaged solar irradiation in Bucharest with the daily and annual values respectively are given below (University of Massachusetts 2004),

$$S = 3.87 \frac{kWh}{m^2 day} = 1414 \frac{kWh}{m^2 year},$$

for a local horizontal surface, under averaged turbidity conditions. From the ESRA database, the value of 3.7 results. In the same database, the optimal irradiation angle is given equal to 35°, although the local latitude is 45°. The difference is coming from the Earth inclination to the ecliptic. As far as the mirror system is optimally controlled, the radiation at the optimal angle must be accounted, as equal to:

$$S = 4.25 \frac{kWh}{m^2 day}, \tag{49}$$

and the mean diurnal insolation time at the same location in Bucharest equal to

$$t_S = 6.121 \frac{h}{day} \tag{50}$$

The following solar irradiation intensity received during the daylight time results

$$Q_B \equiv \frac{S}{t_S} = \frac{4.25}{6.121} \equiv 0.6943 \quad \frac{kW}{m^2} \tag{51}$$

The reflector area, directly facing the Sun results, with the value of

$$A_S \equiv \frac{Q}{Q_B} = \frac{3390}{0.6943} \equiv 4882.4 \quad m^2 \tag{52}$$

Due to different angular positions of the mirrors versus the straight direction to the Sun, due to their individual location on the positioning circle, at least 50% extra reflector area is required to collect the desired radiating power from the Sun, or

$$A_R \equiv 1.5 \cdot A_S = 4882.4 \cdot 1.5 \equiv 7323.6 \, m^2 \tag{53}$$

When 3-m height mirrors are accommodated into circular rows of 200 meters diameter, that means a built surface of 1885 m^2 each, a number of 4 concentric rows must be provided to assure the required solar radiance on the draught tower, or 8 concentric semi-circle rows placed towards the north of the tower. The solution is materialized in Fig. 1. The provided power output must be considered when at least a 40% efficiency of the air turbine is involved, contouring a $2.71 \cdot 0.4 \cong 1\,MW$ real output of the power-plant.

In contrast to the natural gravity air advent, when a turbine or other means of energy extraction are present, the characteristic of the tower suffers a major change however. The tower characteristic includes now the kinetic energy removal by the turbine under the form of externally delivered mechanical work.

5. Turbine effect over the gravity-draught acceleration

The turbine could be inserted after or before the air heater. For practical reasons, the turbine block is better imbedded right upwind the solar receiver (Fig. 9), forcing the raising of the position of the receiver and thus a better insolation of the heater along the whole daylight.

According to the design in Fig. 16, a turbine is introduced in the SEATTLER facility next to the solar receiver, with the role to extract at least a part of the energy recovered from the sun radiation and transmit it to the electric generator, where it is converted to electricity. The heat from the flowing air is thus transformed into mechanical energy with the payoff of a supplementary air rarefaction and cooling in the turbine. The best energy extraction will take place when the air recovers entirely the ambient temperature before the solar heating, although this desire remains for the moment rather hypothetical. To search for the possible amount of energy extraction, the quotient ω is introduced, as further defined. Some differences appear in the theoretical model of the turbine system as compared to the simple gravity draught wind tunnel previously described.

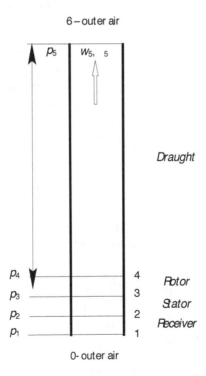

Figure 9. Main stations in the turbine cold-air draught tower.

To describe the model for the air draught with mechanical energy extraction we shall resume some of the formulas from above. First, the process of air acceleration at tower inlet is governed by the same incompressible energy (constant density ρ_0) equation,

$$p_1 = p_0 - \frac{\dot{m}^2}{2\rho_0 A^2} \qquad (54)$$

The air is heated in the solar receiver with the amount of heat q, into a process with dilatation and acceleration of the airflow, accompanied by the usual pressure loss, called sometimes as "*dilatation drag*" (Unger 1988). Considering a constant area cross-section in the heating solar receiver zone of the tube and adopting the variable r for the amount of heating rather than the heat quantity itself (19), with a given value for

$$\beta = \frac{T_1}{T_2} < 1, \qquad (55)$$

the continuity condition shows that the variation of the speed is given by

$$c_2 = c_1 / \beta \qquad (56)$$

No global impulse conservation appears in the tower in this case, as long as the turbine is a source of impulse extraction from the airflow. Consequently the impulse equation will be written for the heating zone only, where the loss of pressure due to the air dilatation occurs,

$$p_2 + \frac{\dot{m}^2}{\rho_2 A^2} = p_1 + \frac{\dot{m}^2}{\rho_0 A^2} - \Delta p_R \qquad (57)$$

A possible pressure loss due to friction into the lamellar solar receiver is considered through Δp_R.

The dilatation drag is thus perfectly identified and the total pressure loss Δp_Σ from outside up to the exit from the solar heater is present in the expression

$$p_2 = p_0 - \frac{\dot{m}^2}{2\rho_0 A^2} - \frac{\dot{m}^2}{\rho_2 A^2} + \frac{\dot{m}^2}{\rho_0 A^2} - \Delta p_R \equiv p_0 - \Delta p_\Sigma \qquad (58)$$

Observing the definition of the rarefaction factor in (54) and using some arrangements the equation (58) gets the simpler form

$$p_2 = p_0 - \frac{\dot{m}^2}{\rho_0 A^2} \cdot \frac{r+1}{2(1-r)} - \Delta p_R \qquad (59)$$

The thermal transform further into the turbine stator grid is considered as isentropic, where the amount of enthalpy of the warm air is given by

$$q = \frac{p_1}{\rho_1} \cdot \frac{1-\beta}{\beta} + \frac{\dot{m}^2}{\rho_1^2 A^2} \cdot \left[\frac{1-\frac{\Gamma}{2}\beta}{\beta} + \frac{\frac{\Gamma}{2}-1}{\beta^2} \right] - \frac{\Delta p_R}{\rho_1} \cdot \frac{1}{\beta}$$

If the simplifying assumption is accepted that, under this aspect only, the heating progresses at constant pressure, then a far much simpler expression for the enthalpy fall in the stator appears,

$$\Delta h_{23} = \omega q = \omega c_p T_2 r \tag{60}$$

To better describe this process a choice between a new rarefaction ratio of densities ρ_3/ρ_2 or the energy quota ω must be engaged and the choice is here made for the later. Into the isentropic stator the known variation of thermal parameters occurs,

$$\frac{T_3}{T_2} = 1 - \omega r, \tag{61}$$

$$\frac{p_3}{p_2} = (1 - \omega r)^{\frac{\kappa}{\kappa-1}}, \tag{62}$$

$$\frac{\rho_3}{\rho_2} = (1 - \omega r)^{\frac{1}{\kappa-1}} \tag{63}$$

The air pressure at stator exit follows from combining (62) and (59) to render

$$p_3 = \left[p_0 - \frac{\dot{m}^2}{\rho_0 A^2} \cdot \frac{r+1}{2(1-r)} - \Delta p_R \right] (1 - \omega r)^{\frac{\kappa}{\kappa-1}} \tag{64}$$

Considering the utilization of a Zölly-type turbine, its rotor wheel keeps thermally neutral by definition and thus no variation in pressure, temperature and density appears in the rotor channel. The only variation is in the direction of the air motion, preserving its kinetic energy as constant.

Thus the absolute velocity of the airflow decreases from the value c_3 to the value $c_3 \sin\alpha_1$ and this kinetic energy variation is converted to mechanical work delivered outside. Consequently $\rho_4 = \rho_3$, $p_4 = p_3$, $T_4 = T_3$ and thus the local velocity at turbine rotor exit is given by

$$c_4 = \frac{c_1}{(1-r)(1-\omega r)^{\frac{1}{\kappa-1}}} \tag{65}$$

The air ascent in the tube is only accompanied by the gravity up-draught effect due to its reduced density, although the temperature could drop to the ambient value. We call this quite strange phenomenon the *cold-air draught*. It is governed by the simple gravity form of Bernoulli's equation of energy,

$$p_5 = p_3 - g\rho_3 \ell \tag{66}$$

The simplification was assumed again that the air density varies insignificantly during the tower ascent. The value for p_3 is here the one in (65). At air exit above the tower a sensible braking of the air occurs in compressible conditions, although the air density suffers insignificant variations during this process.

The energy equation in the form of Bernoulli is used to retrieve the stagnation pressure of the moving air above the upper exit from the tower, under incompressible condition when the density remains constant,

$$p_6{}^* = p_5 - \frac{\Gamma}{2}\rho_5 c_5^2 = p_5 + \frac{\Gamma}{2} \cdot \frac{\dot{m}^2}{\rho_3 A^2} = p_5 + \frac{\Gamma}{2} \cdot \frac{\dot{m}^2}{\rho_0 A^2} \cdot \frac{p_0}{p_3} \tag{67}$$

Value for p_5 from (66) and for the density ratio from (54) and (63) are now used to write the full expression of the stagnation pressure in station "6" as

$$p_6{}^* = (p_0 - \Delta p_R)(1 - \omega r)^{\frac{\kappa}{\kappa-1}} - \frac{\dot{m}^2}{\rho_0 A^2} \cdot \frac{r+1}{2(1-r)} \cdot (1 - \omega r)^{\frac{\kappa}{\kappa-1}} + \frac{\dot{m}^2}{\rho_0 A^2} \cdot \frac{\Gamma}{2} \cdot \frac{1}{(1-r)\cdot(1-\omega r)^{\frac{1}{\kappa-1}}} - g\rho_4 \ell \tag{68}$$

It is observed again that up to this point the entire motion into the tower hangs on the value of the mass flow-rate, yet unknown. The mass flow-rate itself will manifest the value that fulfils now the condition of outside pressure equilibrium, or

$$p_6{}^* = p_0 - g\rho_0 \ell \tag{69}$$

This way the air pressure at the local altitude of the outside atmosphere equals the stagnation pressure of the escaping airflow from the inner tower. Introducing the equation (68) in equation (69), after some re-arrangements of the terms, the dependence of the global mass flow-rate along the tower, when a turbine is inserted after the heater, is given by the developed formula:

$$R^2(\gamma) \equiv \frac{\dot{m}^2}{2g\ell\rho_0^2 A^2} = \frac{1-r}{(r+1)(1-\omega r)^{\frac{\kappa+1}{\kappa-1}} - \Gamma}(1 - \omega r)^{\frac{1}{\kappa-1}} + \frac{p_0}{g\rho_0\ell}\left[(1 - \omega r)^{\frac{\kappa}{\kappa-1}} - 1\right] - \frac{\Delta p_R}{g\rho_0\ell}(1 - \omega r)^{\frac{\kappa}{\kappa-1}} \tag{70}$$

where the notations are again recollected

$r = \dfrac{\rho_0 - \rho_2}{\rho_0}$, the dilatation by heating in the heat exchanger, previously denoted by r;

ω = the part of the received solar energy which could be extracted in the turbine;
Δp_R= pressure loss into the heater and along the entire tube either.

All other variables are already specified in the previous chapters. It is clearly noticed that by zeroing the turbine effect (ω = 0) the formula (70) reduces to the previous form in (37), or by neglecting the friction to (38), which stays as a validity check for the above computations.

For different and given values of the efficiency ω the variation of the mass flow-rate through the tube depends of the rarefaction factor r in a parabolic manner.

6. Discussion on the equations

Notice must be made that the result in (70) is based on the convention (60). The exact expression of the energy q introduced by solar heating yet does not change this result significantly. Regarding the squared mass flow-rate itself in (70), it is obvious that the right hand term of its expression must be positive to allow for real values of R^2. This only happens when the governing terms present the same sign, namely

$$\left\{(r+1)(1-\omega r)^{\frac{\kappa+1}{\kappa-1}} - \Gamma\right\} \cdot \left\{1 - (1-r)(1-\omega r)^{\frac{1}{\kappa-1}} + \frac{p_0}{g\rho_0 \ell}\left[(1-\omega r)^{\frac{\kappa}{\kappa-1}} - 1\right] - \frac{\Delta p_R}{g\rho_0 \ell}(1-\omega r)^{\frac{\kappa}{\kappa-1}}\right\} : 0 \qquad (71)$$

The larger term here is the ratio $p_0/(g\rho_0\ell)$, which always assumes a negative sign, while not vanishing. The conclusion results that the tower should surpass a minimal height for a real R^2 and this minimal height were quite huge. Very reduced values of the efficiency ω should be permitted for acceptably tall solar towers. This behavior is nevertheless altered by the first factor in (71) which is the denominator of (60) and which may vanish in the usual range of rarefaction values r. A sort of thermal resonance appears at those points and the turbine tower works properly well.

7. Discussion on denominator

The expression from the denominator of the formulae (70), which gave the flow reportedly, it can be canceled (becomes 0) for the usual values of the dilatation rapport (ratio) gamma and respectively quota part from energy extracted omega. This strange behavior must be explained. The separate denominator in (72) is,

$$A \equiv \left\{(r+1)(1-\omega r)^{\frac{\kappa+1}{\kappa-1}} - \Gamma\right\} = 0 \qquad (72)$$

The curve of zeros and the zones with opposite signs are:

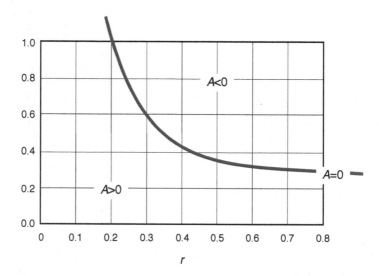

Figure 10. The denominator zeros from (71)

It is yet hard to accept that such a self-amplification or pure resonance of the flow can be real and in fact the formulae (71) does not allow, in its actual form, the geometrical scaling of the tunnel and of the turbine. The rigor of computational formulae is out of any discussion, this showing that the previous result outcomes from the hypotheses adopted. Among those, the hypothesis of isobaric heating before the turbine is obviously the most doubtful.

8. Improved model

Analyzing the simple draught only, observe how easily the hypothesis of isobaric heating leads to an incomplete result, by eliminating the drag produced by the thermal dilatation and the acceleration throw heating, thus reducing the problem to a linear one, without physical anchorage. It could be presumed that the acceptance of relation (57) for the cooling in the stator, relation where it was presumed that the anterior heating performed isobaric, induces an excessive rigidity in the computational model. Replacing this very simple relation between the temperatures and the heat added to the fluid through a non-isobaric relation complicates drastically the model, which becomes completely nonlinear.

It remains to be analyzed whether such an inconvenient model leads to physically acceptable results for the values of mass flow-rate in the turbine tower.

The isobaric relation (60) will be replaced by the exact equation,

$$\Delta h_{23} \equiv c_p T_2 \left(1 - \frac{T_3}{T_2}\right) = \omega q(r),$$ (73)

where the heat received in non isobaric heat exchanger is expressed, through the equation of energy, in the complete form:

$$q(r) = c_p T_0 \frac{r}{1-r} - \left(1 - \frac{\Gamma}{2}\right) \cdot \frac{\gamma}{\Gamma(1-r)^2} \cdot \frac{\dot{m}^2}{\rho_0^2 A^2} - \frac{\Delta p_R}{\Gamma(1-r)\rho_0},$$ (74)

to take also into account the possible pressure losses due to friction in the solar receiver Δp_R.

The absorbed heat (74) will also be used in its complete form in the relation that supplies the pressure at stator exit:

$$\frac{p_3}{p_2} = \left(1 - \omega \frac{q}{c_p T_2}\right)^{\frac{\kappa}{\kappa-1}},$$ (75)

fact that obviously induces another level of non-linearity. Using also the equation of state, the pressure from the stator exits writes from (71),

$$p_3 = p_2^{-\frac{1}{\kappa-1}} \left[p_2 - \omega \left(r\, p_0 - \frac{2-\Gamma}{2} \cdot \frac{r}{1-r} \cdot \frac{\dot{m}^2}{\rho_0 A^2} - \Delta p_R \right) \right]^{\frac{\kappa}{\kappa-1}},$$ (76)

and for the value p_2 the pressure losses from the entrance through Bernoulli acceleration and in the heater will be now respectively inferred,

$$p_2 = p_0 - \frac{\dot{m}^2}{\rho_0 A^2} \left(\frac{\Gamma}{2} + \frac{r}{1-r} \right) - \Delta p_R$$ (77)

Taking into account the draught from the tower (66) and the fluid brake at exit (67), the equilibrium of static pressure reads

$$(p_2 - \omega \beta \chi\, p_0)^{\frac{2}{\kappa-1}} \left[\frac{(p_2 - \omega \beta \chi\, p_0)^\kappa}{g\, \rho_0 \ell} - \beta \right] + \frac{\Gamma}{2} \frac{\dot{m}^2}{\rho_0^2 A^2 g\, \ell \beta}\, p_2^{\frac{2}{\kappa-1}} + (1-\pi)[\, p_2(p_2 - \omega \beta \chi\, p_0)]^{\frac{1}{\kappa-1}} = 0$$ (78)

Here the notation was used:

$$\pi = \frac{p_0}{g\, \rho_0 \ell} >> 1$$ (79)

In the followings the undimensionalised flow-rate D^2 will be considered as the solving variable of the problem, a variable that naturally appears from the previous equation (78), under the form of the ratio

$$D^2 \equiv \frac{R^2}{\pi} = \frac{\dot{m}^2}{2\rho_0^2 A^2 g \ell} \cdot \frac{8 \rho_0 \ell}{p_0} = \frac{c_1^2}{2 R T_0} \equiv \left(\frac{c_1}{c_0}\right)^2, \tag{80}$$

where also naturally appears the characteristic velocity c_0 of the air c_0, namely

$$c_0 \equiv \sqrt{2 R T_0} \approx 415,5 \, m/s \tag{81}$$

The characteristic velocity c_0 is actually related to the local sound velocity in the air a_0, manifesting proportional to it, so that the relative mass flow-rate can be written in the absolutely equivalent form,

$$a_0 \equiv \sqrt{\kappa R T_0} \approx 348,2 \, m/s \tag{82}$$

in connection with which the relative flow-rate couls also be expressed, in the form

$$D^2 \equiv \frac{R^2}{\pi} = \frac{\kappa}{2} \frac{c_1^2}{a_0^2} = \frac{\kappa}{2} M_1^2, \tag{83}$$

in other words this flow-rate is proportional to the squared local Mach number of the flow.

From (78) the equation of the flow-rate D^2 is obtained as a function of the working conditions, expressed through the parameters ω and r,

$$(a \cdot c - b\, D^2)^5 [(c - b\, D^2)^{1,4} - c^{2,4}] + d \cdot c^{0,4} D^2 (c - e\, D^2)^5 + f \cdot c^{1,4} (c - e\, D^2)^{2,5} (a \cdot c - b\, D^2)^{2,5} = 0 \tag{84}$$

where the constant coefficients are again reproducing those working conditions,

$$\begin{aligned}
a &\equiv 1 - \omega r, & e &\equiv 2r + \Gamma(1-r), \\
c &\equiv 1 - r, & d &\equiv \Gamma \pi, \\
f &\equiv 1 - \pi, & b &\equiv e - \omega r\,(2 - \Gamma).
\end{aligned} \tag{85}$$

The algebraic, non linear equation (80) is now solved using a standard numerical method to obtain solutions for the mass flow-rate, as depending on the different working conditions concerning the heating level applied in the solar receiver r and respectively the degree of recovery of the heat introduced through the receiver ω. For a complete recovery of energy ($\omega=1$), the numerical solutions are given in the following table (Table 3):

It proves however that the above given model is not properly reproducing the Stack-Turbine (S-T) characteristic at low heating rates ($r \to 0$), while at the upper end ($r \to 1$) it acceptably does this. The same improper behavior is observed when for example a compressible, variable density acceleration at the stack entrance is considered in the simple draught. In that case the "false" equation appears,

$$\frac{\dot{m}^2}{2g\ell\,\rho_0^2 A^2} = 1 - r, \qquad (86)$$

or

$$D^2 - (1-r)/\pi = 0 \qquad (87)$$

A very slight change in the assumptions could therefore deeply affect the result of the simulation modeling, due to the small overall magnitudes of pressure and density gradients along the S-T channel.

r	D²
0	3,50
0,1	-
0,2	(1,280)
0,3	(0,875)
0,4	0,611331000
0,5	0,428261326
0.6	0,298397500
0,7	0,199248700
0,8	0,1098315 și 0,012898027
0,9	0,0634500 și 0,055130000
1,0	0,00

Table 3. The equilibrium flow-rate as a function of the rarefaction y for ω=1

The results are plotted in the diagram from Fig. 11. The discharge characteristic of the tunnel resulting from the given assumptions is drawn in dark red.

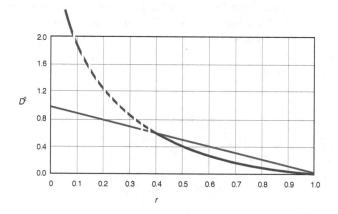

Figure 11. Discharge characteristic of SEATTLER tower.

9. Energy output of the gravity-draught accelerator

The main concern and reluctance for the classical solar towers comes from the regular perception that the energetic efficiency of those systems is unsatisfactory. Largely correct, this perception does not further stand valid for gravity draught towers and to prove this a piece of attention must be allocated to the energy balance.

The equation of energy in its rough form (3) needs thus further attention. Pointing the values to the exit station "2" of the receiver (Fig. 2) we first observe that the gain in kinetic energy e_g given by the tower directly, per kilogram of air, defined by

$$e_g \equiv \frac{w_2^2}{2}$$

is equal to

$$e_g = q - g(z_2 - z_1) - c_p(T_2 - T_0) \tag{88}$$

where the first right-hand term is the total heat introduced into the stack per one kg of air, and the last term represents the heat consumed for directly heating the air to the final temperature T_2. The second term, which acts as a reducer of the efficiency, is relatively small in comparison to the others.

The quantity of kinetic energy transferred to the air is the difference that remains available. This entire amount could be used to produce energy, without any thermal or mechanical loss. Physically, the heat introduced in the air to create the up-draught along the tower could entirely be extracted into useful mechanical work through a low temperature wind turbine, and the draught is maintained due to the low air density despite the energy extraction in the tower.

The process remains however greatly dependent to the optimal selection of the heating level and of the utilization of the solar radiation in an efficient manner. The problem with the cloudy weather and the energy stocking during the night are solved through heat accumulators of specific construction.

10. Conclusion

The principle of a solar energy power plant, based on a mirror-type collector, is depicted in the nearby drawing. It represents the application of the WINNDER thermal accelerator principle into the ecological and sustainable means of accelerating the air without any moving device and, consequently, with a very low noise and turbulence level, ideal for aeroacoustic applications. A multiple-rows array of controllable ground mirrors are installed around. In this manner a highly efficient utilization of the solar energy is available, due to the known high release coefficient of the mirror surfaces. Means to follow the Sun along its apparent trajectory are common and available at low cost today. Problems regarding the maintenance of the system can be solved through a proper technological and economic management of the facility.

It does not seem however equally attractive for energy production, despite the clean method involved, but this represents a first sight impression, easily dismounted through an in-depth analysis. The computational model depicted above shows that the resources for producing energy trough the solar gravity draught are high enough and represent an interesting resource of green energy of a new and yet unexplored type.

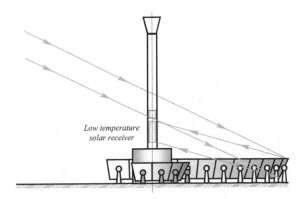

Low temperature solar receiver

Figure 12. Principle of WINNDER concentrator for aeroacoustic applications.

Although the equipment costs of the present project are much higher than the Green-house power plant ones, it is believed that the overall costs are still competitive and the proposed solution of reflector tower is useful. One of the explanations resides in the fact that the reflexivity of the mirrors is very high. The design example given above high-lights the main factors.

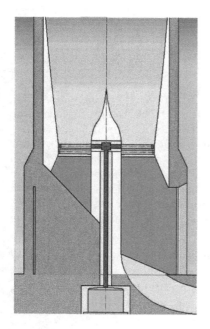
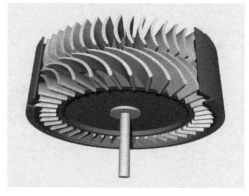

Figure 13. STRAND air turbine.

This example shows that a circular ground surface of roughly 0.8 *ha* at maximum must be used to produce a 1 *MW* power output, or 0.8 *ha/MW*. The figure is to be compared to the one of Manzanares power-plant in Spain, built under the solar green-house collector pro-gram, where the amount of occupied soil equals 90 *ha/MW* or 100 times more (peak output of 50 *kW* for a collector diameter of 240 meters). This capacity is also higher than the surface-to-power production intensity of photoelectric cells of 1.0 *ha/MW* (Energy Form EIA-63B). The costs of Solar cells of 4.56 *$/PeakW* (1995) are still high.

After the data in (Schleich et al. 2005) this value equals 0,94 and this adds to the very high absorbing properties of the tower walls. It serves here as a nice illustration of possible extra applications of the chimney draught effects in directly producing electrical power.

As another comparison item, the newly renovated Solar Two solar thermal electric generat-ing station, located in California's Mojave Desert, consists of 1,900 motorized mirrors sur-rounding a generating station with 10 megawatts of capacity, which began operation in early 1996. It is part of an effort to build a commercially viable 100-*MW* solar thermal system by 2000 (Energy 1997). The 10-MW Solar Two solar thermal electric plant near Barstow, CA, began operation in early 1996 on the site of the Solar One plant. Solar Two differs from Solar One primarily in that it includes a molten-salt storage system, which allows for several hours of base-load power generation when the sun is not shining.

The molten salt (an environmentally benign combination of sodium nitrate and potassium nitrate) allows a summer capacity factor as high as 60%, compared with 25% without storage. The plant consists of 1,926 motorized mirrors focused on a 300-*ft*-high central receiver generating station rated at 10 *MW*. Molten salt from the "cold" salt tank (at 550ºF) is heated to 1,050ºF and stored in the "hot" salt tank. Later the hot salt is passed through a steam generator to produce steam for a conventional steam turbine.

Equipment costs of WINNDER are higher than for the Greenhouse power plants, still the overall costs of exploitation and maintenance are competitive and the proposed combination of mirror array and draught tower is literally efficient. It remains to convince the investors of the efficiency of this exotic energy producer.

The gravitational up-draught due to Archimedes's effect does not contribute, in any way, to the balance of energy. It simply remains the driver of the air into the stack and the solar energy introduced in the system is the only source of air acceleration and further production of electric energy within a turbo-generator. Consequently it does not seem specifically attractive for energy production, although it provides the cleanest energy ever and involves the lowest levels of losses.

Author details

Radu D. Rugescu*

Address all correspondence to: rugescu@yahoo.com

University "Politehnica" of Bucharest, Romania

References

[1] Bejan, A., Convection Heat Transfer, New York, Wiley and Sons, 1984.

[2] Cirligeanu, R., Rugescu, R. D., and AlinaBogoi (2010), TRANSIT code for turbine flows in solar-gravity draught power plants, Paper GT2010-22518, Proceedings of the ASME International Conference on Gas Turbines, June 14-18, 2010, Glasgow, Skotland, UK.

[3] Gannon, A. J., T. W. Von Backström, Solar Chimney Turbine Performance, ASME Journal of Solar Energy Engineering, 125, 2003, p101-106.

[4] Haaf, W., Solar Chimneys, Part II: Preliminary Test Results from the Manzanares Pilot Plant, Int. J. Sol. Energy, 2 (1984), pp. 141–161.

[5] Jaluria, Y., Natural Convection, Heat and Mass Transfer, Oxford, New York, Pergamon Press, 1980.

[6] Mancini, Th. R., Solar-Electric Dish Stirling System Development, SAND-97-2924C, 1998.

[7] Mueller R.W., Dagestad K.F., Ineichen P., Schroedter M., Cros S., Dumortier D., Kuhlemann R., Olseth J.A., Piernavieja G., Reise C., Wald L., Heinnemann D. (2004), Rethinking satellite based solar irradiance modelling - The SOLIS clear sky module. Remote Sensing of Environment, 91, 160-174.

[8] Raiss, W., Heiz- und Klimatechnik, Springer, Berlin, vol. 1, pp. 180-188, 1970.

[9] Rugescu, R. D., ThermischeTurbomaschinen, ISBN 973-30-1846-5, E. D-P. Bucharest 2005.

[10] Rugescu, R. D., T. G. Chiciudean, A. C. Toma, F. Tache, Thermal Draught Driver Concept and Theory as a Tool for Advanced Infra-Turbulence Aerodynamics, in DAAAM Scientific Book 2005, ISBN 3-901509-43-7 (Ed. B. Katalinic), DAAAM International Viena, 2005.

[11] Rugescu, R. D., D. A. Tsahalis, and V. Silivestru (2006), Solar-Gravity Power Plant Modeling Uncertainties, WSEAS Transactions on Power Systems, ISSN 1790-5060, 10(2006), pp.1713-1720.

[12] Rugescu, R. D. (2008), "Technology of CFD in space engines and solar-gravitational power plants", International Journal on Energy Technology and Policies, ISSN 1472-8923, On-line ISSN 1741-508X, 6, 1-2(2008), p.124-142.

[13] Rugescu, R. D., Silivestru V., and Ionescu M. D. (2008),Technology of Thermal Receivers and Heat Transfer for SEATTLER towers, Chapter 59 in the DAAAM International Scientific Book 2008, pp. 697-742, B. Katalinic (Ed.), published by DAAAM International Vienna, ISBN 3-901509-69-0, ISSN 1726-9687, Vienna, Austria.

[14] Rugescu, D. R. and R. D. Rugescu (2009), Chapter 1 "Potential of the solar energy on Mars", pp. 01-24, in Rugescu, D. R. (editor), "Solar Energy", ISBN 978-953-307-052-0, Ed. INTECH, Rjeka, Croatia, 2010, 462 p.

[15] Rugescu, R. D., R. B. Cathcart, Dragos R. Rugescu, and S. P. Vataman (2010), Electricity and Freshwater Macro-project in the Arid African Landscape of Djibouti, AR-GEO-C3 Conference, Exploring and harnessing the renewable and promising geothermal energy, 22-25 November, Djibouti.

[16] Rugescu, R. D., R. B. Cathcart, and Dragos R. Rugescu, Electricity and Freshwater Macro-project in an Arid, Ochre African Landscape: Lac Assal (Djibouti), Eighth International Conference on Structural Dynamics, Proceedings EURODYN-2011, Leuven, Belgium, 4-6 July 2011 (pp NA).

[17] Rugescu, R. D., A. Bogoi, and R. Cirligeanu (2011), Intricacy of the TRANSIT manifold concept paid-off by computational accuracy, Proceedings of ICMERA-2011 Conference, ISBN 978-981-07-0420-9, ISSN 2010-460X, Bucharest 20-22 October, Romania, pp. 61-65.

[18] Rugescu, R. D. (Ed.), Solar Power, ISBN 978-953-51-0014-0, INTECH Press, Rjeka, Croatia, Hard cover, 378 pages, Feb. 2012.

[19] Rugescu, R. D., and C. Dumitrache (2012), Solar Draught Tower Simulation and Proof by Numerical-Analytical Methods, Paper V&V2012-6191, ASME 2012 Verification and Validation Symposium, Las Vegas, May 2-4, 2012, p. 73.

[20] Rugescu, R. D., Demos T. Tsahalis, CiprianDumitrache (2012), Extremely High Draught Tower Efficiency Proof by Conservation of Energy, Invited plenary lecture, Proceedings of the 5th IC-SCCE 2012 From Scientific Computing to Computational Engineering, 4-7th July 2012, Athens, Greece, pp. 243-248.

[21] Scharmer, K., Greif, J. (2000), The European Solar Radiation Atlas, Presses de l'Ecole des Mines, Paris, France.

[22] Schiel, W., T. Keck, J. Kern, and A. Schweitzer, Long Term Testing of Three 9 kW Dish-Stirling Systems, ASME International Solar Energy Conference, San Francisco, CA, USA, March 1994.

[23] SchleichBergermann und Partners (2005), EuroDish System Description.

[24] Tache, F., Rugescu, R. D., Slavu, B., Chiciudean, T. G., Toma, A. C., and Galan, V. (2006), Experimental demonstrator of the draught driver for infra-turbulence aerodynamics, Proceedings of the 17th International DAAAM Symposium, "Intelligent Manufacturing & Automation: Focus on Mechatronics & Robotics", Vienna, 8-11th November 2006.

[25] Unger, J., Konvektionsströmungen, B. G. Teubner, ISBN 3-519-03033-0, Stuttgart, 1988.

[26] Energy Information Administration, Form EIA-63B, "Annual Photovoltaic Module/ Cell Manufacturers Survey."

[27] Energy Information Administration, Office of Coal, Nuclear, Electric and Alternate Fuels, Renewable Energy Annual 1996, U.S. Department of Energy, Washington, DC 20585, April 1997.

[28] University of Massachusetts (2004), Lowell Photovoltaic Program, International Solar Irradiation Database, Version 1.0, Monthly solar irradiation.

Chapter 4

Thin Film Solar Cells:
Modeling, Obtaining and Applications

P.P. Horley, L. Licea Jiménez, S.A. Pérez García,
J. Álvarez Quintana, Yu.V. Vorobiev, R. Ramírez Bon,
V.P. Makhniy and J. González Hernández

Additional information is available at the end of the chapter

1. Introduction

The development of human civilization was fueled by different energy sources throughout its history, with the past decades clearly showing a trend of using environment-friendly energy. Among the energy sources available from the Nature, solar energy has a special place [1, 2]. It is available in vast quantities, especially in countries with high insolation level such as México. Transformation of solar light into electricity takes place in photovoltaic devices – solar cells – that do not require much maintenance throughout their operation cycle and can function as stand-alone devices allowing the use of electrical equipment even in most remote areas. The energy produced by the solar cells during the day can be stored in accumulators and used during the night, making solar-powered equipment practically self-sustainable if the required number of sunlight hours per day is available.

Currently, most commercial solar cells are made of silicon due to its vast availability and silicon technology that reached the state of perfection, allowing to achieve the conversion efficiency of almost 28% (single silicon cell, [3]). Actually, world production of photovoltaics is dominated by polycrystalline silicon cells representing 94% of the market [4]. These devices based on silicon wafers are called the "first generation" of photovoltaic technology. Monocrystalline silicon devices are effective but expensive. Similarly, solar cells based on other semiconductor materials, called "second generation", such as group II-VI and III-V heterostructures, are capable of efficiencies over 40% (GaInP/GaAs/GaInNAs, [3]) but depend significantly on the quality of the junction that may contain defects acting as effective recombination centers, reducing considerably the concentration of photo-generated carriers.

Finally, there is the "third generation" of photovoltaic devices that embraces solar cells based on organic semiconductors. These materials usually afford moderate efficiency – about 11% (dye-sensitized cells [3]); however, they are cheap and easy to obtain, which is a very attractive point for industrial-scale production. Also, organic materials can be deposited on flexible substrates, widening the spectrum of their possible applications.

In this chapter, we present the results for several types of heterojunction solar cells that are particularly focused on the use of thin film devices for photovoltaic conversion [5]. We discuss the benefits of computer simulations for improvement of AlGaAs/GaAs solar cells, suggesting the optimal values of aluminum contents and thickness of the window layer. We propose the isovalent substitution method as a promising technological approach for crafting near-to-perfect junction boundary with reduced mismatch of lattice parameters and thermal expansion coefficients, illustrating it for the case of CdTe/CdS heterostructures. Aiming to lower the cost of solar cell production, we consider the option of chemical bath deposition for CdS/PbS solar cell, proposing environment-friendly variation that significantly reduces (and even disposes of) the use of toxic ammonia that is characteristic for a common chemical bath deposition process for CdS films. We also address the question of organic solar cells, discussing the mechanisms of current transport in a cell based on poly (3-hexylthiophene). Finally, we consider the question of excess heating that is characteristic to the photovoltaic devices (especially those operating under concentrated light conditions), proposing to use the experience gained from nano-thermoelectric formations used to remove the extra heat from the devices composing microchips.

2. General theoretical modeling

Let us consider the basic physical processes taking place in a semiconductor solar cell with a heterojunction (Fig. 1). The device is composed by two semiconductors with different band gap values [1, 6]. The wider-band material forms so-called window layer (for which the corresponding characteristics in Fig. 1 have the subscript "W") and is used to process high-energy photons, allowing low-energy photons to pass through. These became absorbed in the narrower-band material forming the absorber layer (hence the subscript "A" in Fig. 1). The thickness of the corresponding layers will be referred to as D_W and D_A, correspondingly, making the total device thickness equal to $D_{WA} = D_W + D_A$. The presence of window and absorber layers allows to optimize solar spectrum use and reduce device heating that is more prominent in p-n junctions, where more absorbed photons have the energies exceeding band gap of the material. The other benefit consists in increase of material choice for creating the junction, because not all semiconductors can be obtained in both modifications with p- and n- conductivity. On the negative side, the mismatch of lattice parameters of junction components create undesirable defects, and difference in thermal expansion coefficient may be critical for stability of the device if used under the elevated temperatures.

The contact of two materials with different conductivity type leads to the formation of space charge region [7] associated with diffusion potential difference U_d. Within this so called depletion region of width $w_n + w_p$ (Fig. 1) energy band bending occurs. In general case, due to the difference of band gap values of window and absorber layers (E_{gW} and E_{gA}, respectively) there will be band discontinuities ΔE_V and ΔE_C, introducing additional energy barriers for the carriers and paving the way for different types of tunneling effects. The band diagram of a heterojunction can be constructed using electron affinities for both materials – χ_W and χ_A, respectively, which allows to construct band diagram of the structure.

Under illumination, the electrons obtain the energy sufficient for moving into conduction band, creating holes in the valence band. The resulting non-equilibrium electron-hole pair can disappear due to recombination. However, if it is generated in the vicinity of the junction, the embedded electric field of the space charge region will exert different forces on the carriers in accordance with their charges, moving them towards the contacts where they contribute to the photo-current of the external circuit.

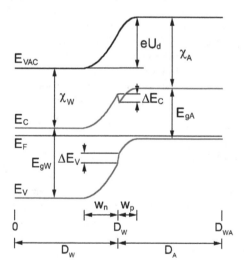

Figure 1. Band diagram of a heterojunction with main parameters denoted.

The transport of the carriers, in addition to the action of the embedded electric field, is also governed by the diffusion caused by the difference in concentrations of electrons and holes in the corresponding parts of the device. This mechanism can be described as [8]

$$J_n = en(x)\mu_n E(x) + eD_n \frac{dn}{dx} \tag{1}$$

$$J_p = ep(x)\mu_p E(x) - eD_p \frac{dp}{dx} \tag{2}$$

where n, p are concentration of electrons and holes, μ_n and μ_p are their mobilities and D_n, D_p are diffusion coefficients. The action of the electric field is limited to the space charge region that is characterized by the thickness

$$w_n + w_p = \sqrt{\frac{2\varepsilon_W\varepsilon_A N_A(U_d - U)}{eN_D(\varepsilon_W N_D + \varepsilon_A N_A)}} + \sqrt{\frac{2\varepsilon_W\varepsilon_A N_D(U_d - U)}{eN_A(\varepsilon_W N_D + \varepsilon_A N_A)}} \tag{3}$$

with dielectric constants of window and absorber materials ε_W and ε_A, respectively. The values of N_D and N_A correspond to the concentration of donors and acceptors that define conductivity type of the materials forming heterojunction. The height of the energy barrier, U_d, can be manipulated by application of a voltage U, which is also included into (3). Calculating the value of $w_n + w_p$ it can be shown that the space charge region usually is of negligible thickness in comparison with that of the entire device. Therefore, one can simplify the equations (1) and (2) by keeping only diffusion terms:

$$J_n = eD_n \frac{dn}{dx}$$
$$J_p = -eD_p \frac{dp}{dx} \tag{4}$$

The resulting expressions can be rewritten relating current variation to the difference of recombination and generation rates. For simplicity, we will present here only equations describing window layer, for which the minority carriers are holes:

$$D_p \frac{d^2 p}{dx^2} = r_p - g_W \tag{5}$$

Recombination rate $r_p = (p - p_{n0})/\tau_{pW}$ depends on the difference between non-equilibrium and equilibrium concentrations of holes given in numerator, and a characteristic time τ_{pW} defined by recombination processes taking place in the system. Generation rate $g_W = \alpha_W \Phi_0 \exp(-\alpha_W x)$ includes absorption coefficient of the material α_W, spectral power of incident light flux Φ_0 and distance from the surface x. Substituting these expressions into formula (5) one will obtain

$$\frac{d^2 p}{dx^2} = \frac{p - p_{n0}}{L_p^2} - \frac{\Phi_0 \alpha_W}{D_p} e^{-\alpha_W x} \tag{6}$$

where $L_p = \sqrt{D_p \tau_p}$ is the diffusion length for the holes. Equation (6) should be solved taking into account the boundary conditions. On the front of the cell, the variation of hole concentration is connected with surface recombination rate s_p

$$\left.\frac{dp}{dx}\right|_{x=0} = \frac{s_p}{D_p}(p(0) - p_{n0}) \tag{7}$$

At the boundary with space charge region, the concentration of holes is equal to:

$$p(D_W - w_n) = p_{n0} exp\ (eU\ /\ k_B T) \tag{8}$$

where $k_B T$ is a product of the Boltzmann constant and the temperature. The solution of the equation (6) is usually written in the form

$$p(x) = p_{n0} + C_0^p e^{-\alpha_W x} + A_p e^{x/L_p} + B_p e^{-x/L_p} \tag{9}$$

with $C_0^p = \frac{\alpha_W \Phi_0 L_p^2}{D_p(1 - \alpha_W^2 L_p^2)}$ and coefficients A_p and B_p that can be found from boundary conditions. The similar equations can be obtained for electrons as minority carrier in absorber part of the device, yielding the general solution for carrier concentration as

$$n(x) = n_{p0} + C_0^n e^{-\alpha_W(x - D_W)} + A_n e^{x/L_n} + B_n e^{-x/L_n} \tag{10}$$

where D_W appearing in the first exponent denotes the decrease of light flux upon passing through the window layer. Now, the total current through passing to the external circuit can be obtained as the sum of electron and hole currents at the contacts together with total photo-generation current reduced by integral describing recombination losses:

$$J = J_p(0) + J_n(D_{WA}) + e \int_0^{D_{WA}} g(x)dx - e \int_0^{D_{WA}} r(x)dx. \tag{11}$$

Two first terms in (11) can be easily found using the expressions for carrier concentrations (9), (10) and their relation to the corresponding current components (4):

$$J_p(0) = eD_p \left[\alpha_W C_0^p + \frac{1}{L_p}(B_p - A_p) \right] \tag{12}$$

$$J_n(D_{WA}) = -eD_n \left[\alpha_A C_0^n e^{-\alpha_A D_A} + \frac{1}{L_n}(B_n e^{-D_{WA}/L_n} - A_n e^{D_{WA}/L_n}) \right] \tag{13}$$

The generation term from (11) can be calculated analytically:

$$e \int_0^{D_{WA}} g(x)dx = \Phi_0 \alpha_w e \int_0^{D_W} e^{-\alpha_w x} dx + \alpha_A \Phi_0 e^{-\alpha_w D_W} e \int_0^{D_A} e^{-\alpha_w(x-D_W)} dx$$
$$= \Phi_0 e \left[1 - e^{-(\alpha_w D_W + \alpha_A D_A)} \right] \tag{14}$$

The recombination term is calculated numerically, taking into account the distribution of non-equilibrium carrier concentration in window/absorber and various recombination mechanisms such as direct recombination, Hall-Shockley-Read recombination involving impurity levels in band gap, and Auger recombination for high-energy carriers that transfer their excess energy to another particle [8].

For numerical simulations we considered the heterojunction $Al_xGa_{1-x}As/GaAs$ [9] characterized by a small lattice mismatch of 0.127%. The window layer remains direct-band semiconductor for aluminium contents less than 45%. Material parameters used in our simulations are listed in Table 1 as functions of aluminium content x and temperature T.

Parameter	Value or calculation formula
Dielectric constant ε	$12.90 - 2.84x$
Electronic affinity χ, eV	$4.07 - 1.1x$
DOS effective masses	$m_n^* = (0.067 + 0.083x) m_0; m_p^* = (0.62 + 0.14x) m_0$
Electron mobility μ_n, cm²/(Vs)	$(8 - 22x + 10x^2) \cdot 10^3$
Hole mobility μ_p, cm²/(Vs)	$(3.7 - 9.7x + 7.4x^2) \cdot 10^2$
Band gap E_g, eV	$1.424 + 1.247x - (5.4 \cdot 10^{-4}T^2)/(T+204[K])$

Table 1. Parameters of $Al_xGa_{1-x}As$ used in calculations

In the framework of the current chapter, we studied the dependence of window layer thickness D_W on the efficiency of AlGaAs/GaAs solar cell. All calculations were done for AM1.5 illumination [10]. First, we considered the question about the thickness ratio of window/absorber layers (Fig. 2). The resulting plot has roughly triangular shape with the grayed out area in the bottom left corner where the system is not converging to any solution. As one can see from the figure, the efficiency η exceeding 24% is obtained for thicker cell ($D_{WA} = 300$ μm), which is expected because the junction should have enough material for a considerable absorption of solar light. When the cell is 3 μm thick, the value of η reaches 20% at most. It is also clear from the figure that the cell performs better with a thin window layer. For example, the efficiency over 20% is reachable for a thick cell with window layer thickness under 0.5% of D_{WA}, i.e., $D_W < 1.5$ μm. This result proves that the embedded electric field of space charge region should be located quite close to the surface where the major photo-generation of non-equilibrium carriers takes place, ensuring efficient separation of electron-hole pairs and reducing losses due to the recombination processes. If the junction is located deeper into the cell, the embedded field becomes less efficient. Also, thicker absorber layer augments the number of processed photons that increases the current flowing to the external circuit.

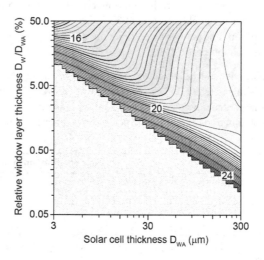

Figure 2. Dependence of $Al_{0.29}Ga_{0.71}As/GaAs$ solar cell performance as a function of window and cell thickness

Another important point is the adjustment of band gap difference between the heterojunction components that modifies the percentage of light processed by window and absorber layers. Having in mind that thicker cell has better overall performance, we performed calculations varying the aluminum contents x in $Al_xGa_{1-x}As$ and thickness of the window layer. These results are presented in Fig. 3. Similarly to Fig. 2, the case of ultrathin window precludes numerical convergence and is greyed out. As one can see from the figure, the efficiency landscape has two prominent details. For comparative thick window layer with D_W above a micron, it has a pronounced maximum at $x = 29.5\%$ that does not shift with variation of D_W by two orders of magnitude. The maximum changes into a wide plateau with $\eta > 17\%$ for x exceeding 20%, following with a quick drop of efficiency for decreasing aluminum content. Increase of x above 30% also causes abrupt drop of the efficiency. We explain this behavior by optimal adjustment of band gaps E_{gW} and E_{gA} ensuring good separation of solar spectra into "high" and "low"-energy parts processed by window and absorber layer, respectively, for 20% $< x <$ 30%. For lower x, the difference of band gaps is insufficient. For higher x, the difference is too big so that the energy of the photons passing the window layer is high in comparison with E_{gA}, which will result in excess Auger recombination rate and increased cell heating.

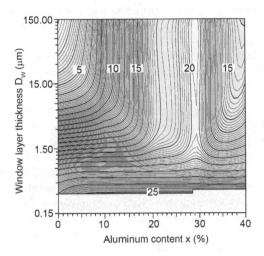

Figure 3. Efficiency of Al$_x$Ga$_{1-x}$As/GaAs solar cell as a function of window layer thickness D_W and aluminum contents x

However, when window layer becomes very thin ($D_W < 1\mu m$), the system starts to behave differently. Now, the generation of the carriers in direct vicinity of junction boundary provides a significant benefit by quick and efficient separation of carriers by electric field associated with space charge region, reducing recombination losses. Nevertheless, even the contour line for $\eta = 23\%$ shows that the cell performs slightly better when aluminum contents gravitates towards $x = 30\%$.

Therefore, theoretical treatment of semiconductor solar cell followed by numerical simulations allowed to obtain useful information about the system, which can significantly simplify the experimental optimization of solar cell parameters by suggesting the most promising ranges of parameters that corresponds to the highest efficiency of photovoltaic conversion of solar energy.

3. Solar cells with CdTe layers grown by isovalent substitution

Among the materials used for solid-state cells, cadmium telluride occupies a special place due to its near-optimal band width of 1.5 eV at 300 K. Direct band of CdTe favors manufacture of thin-film barrier structures [5, 6]. However, despite it is possible to grow CdTe with n- and p-conductivity, the p-n junction solar cells of cadmium telluride are impractical due to high absorption and recombination. Schottky barriers are also not quite useful because many metals form comparatively low barriers with p- and n-type cadmium telluride [11]. Under these conditions, the major flexibility in device design can be attained for heterojunctions, among which the most prominent are thin-film structures of p-CdTe/n-CdS. It is quite easy to obtain solar cells with efficiency of $\eta \approx 16\%$, but much device optimization work is required to achieve the theoretical performance limit of 28% [6]. One of the main reasons reducing the efficiency of solar cell is a large concentration of defects N_S at the junction boundary caused by a mismatch of crystalline and thermal parameters of device components [12]. Therefore, it is important to search for the best material especially for wide-band window layer and improve the reproducibility of technology aiming to create heterostructures with a perfect junction boundary.

One of the promising approaches to solve this problem involves the method of isovalent substitution (IVS) [13], offering considerable advantages over the traditional methods of heterojunction manufacturing [5, 12]. The substituted heterolayers grows down into the substrate, which defines and stabilizes the crystalline structure of the layer. Thus, IVS is used to obtain stable layers of materials with crystalline modifications that do not exist in bulk form. The layers of intermediate solid solutions relax the difference of lattice parameters and thermal expansion coefficients, ensuring low Defect concentration at the junction boundary. Finally, residual base substrate atoms act as isovalent impurities, significantly increasing the temperature and radiation stability of the material [14]. This section reports successful use of isovalent substitution method for formation of heterojunctions of CdTe with wide-band II-VI compounds such as CdS, ZnTe and ZnSe.

The base substrates with the size of 4×4×1 mm^3 were cut from bulk CdS, ZnTe and ZnSe crystals grown by Bridgman method from the stoichiometric melt. At the room temperature the substrates of CdS and ZnSe had n-type conductivity; ZnTe samples were of p-type due to presence of intrinsic point defects. After mechanical polishing the plates were etched chemically in the solution of Cr_2O_3 : HCl = 2 : 3, rinsed in distilled water and dried. These operations ensured mirror-reflective surface of the substrates and bulk luminescence within the corresponding spectral ranges. The heterostructures were formed by annealing of the substrates in quartz containers pumped out to 10^{-4} Torr at the temperatures of T_A=800 –1000 K. The additional charges loaded into containers are listed in Table 2.

Substrate	n-CdS	p-ZnTe	n-ZnSe
Charge	Charge of CdTe, Te and $LiCO_3$ salt	Charge of CdTe and Cd	Charge of CdTe

Table 2. Loads to the containers required to form heterostructures of corresponding type

Annealing process leads to formation of cadmium telluride layer on top of the base substrates, which was verified by spectra of optical reflectivity and transmittivity. The thickness of CdTe layers is controlled by deposition conditions and film conductivity type results to be opposite to that of the substrate. To make diode structures, we polished off the CdTe layer from one side down to the substrate. Ohmic contacts were deposited by melting in In for n-type material and vacuum-sputtering of Ni for the p-type one. The sketch of resulting p-CdTe/n-CdS structure is given in the inset to Fig. 4.

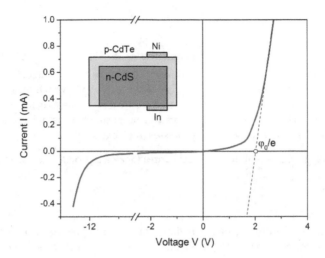

Figure 4. Typical CVC of p-CdTe/n-CdS heterojunction at 300 K. The inset shows a schematic view of the structure.

The heterojunctions studied featured pronounced diode characteristics with rectification co-efficient above 10^4 at 300 K and voltage of 1 V. The typical dark current-voltage curve (CVC) of an example p-CdTe/n-CdS heterojunction is given in Fig. 4. The potential barrier height φ_0 can be estimated extrapolating the straight branch of the curve towards the intersection with the voltage axis (Fig. 4). The value of φ_0 depends on heterojunction type and technological condition used to obtain it, with the maximum barrier values listed in Table 3.

Heterojunction	p-CdTe/n-CdS	n-CdTe/p-ZnTe	p-CdTe/n-ZnSe
φ_0, eV	1.25	1.4	1.3
V_{oc}, V	0.6	0.8	0.7

Table 3. Maximum barrier height and open circuit voltage of heterojunctions at 300 K

Our analysis shows that current-voltage curves at lower bias obey the expression that corresponds to dominating carrier recombination in the space charge region [6]

$$I_{gr} \approx I_{gr}^0 \exp(eV / 2kT) \tag{15}$$

with intercept current I_{gr}^0 obtained for $V=0$. Energy slope of $I_{gr}^0(T)$ curve plotted in axis frame $\ln\left(I_{gr}^0 - 1/T\right)$ is about 1.6-1.7 eV, which agrees with the bandgap of cadmium telluride at 0 K, proving that the space charge region is mainly localized in CdTe and recombination takes part in the thinner component of the junction. Under the higher bias the formula (1) transforms into less steep dependence [5

$$I_{grt} \approx I^0_{grt} \exp(\alpha V + \beta T) \qquad (16)$$

where I^0_{grt} is the intercept current and parameters $\alpha = (5-15)$ V^{-1}, $\beta = (0.01-0.02)$ K^{-1} are independent on both voltage and temperature. For the bias approaching φ_0/e, current transport in the device becomes dominated by over-barrier current I_d, which for the heterojunctions p-CdTe/n-CdS and p-CdTe/n-ZnSe is provided by electrons and for n-CdTe/p-ZnTe diodes – by holes. The inverse current in obtained heterojunctions is determined by carrier tunneling (low bias case) and avalanche processes (high bias case).

The current voltage curve of illuminated structures for $eV \geq 3kT$ obeys expression

$$I_{SC} = I^0_{SC} \exp(eV_{OC}/2kT) \qquad (17)$$

with short circuit current I_{SC} and open circuit voltage V_{OC}. The presence of two in denominator of the exponent corresponds to the major carrier generation in the space charge region. Thermal dependence of the intercept I^0_{SC} photocurrent value is determined mainly by the exponential coefficient $\exp(-E_g/2kT)$. The band gap value E_g appearing here characterizes the material in which the most intensive generation takes place. In the heterostructures synthesized under low T_A the value of E_g is about 1.6 eV, which correspond to the band gap of CdTe at 0 K. At higher synthesis temperature E_g increases up to 1.8-2.0 eV, suggesting that photogeneration of the carriers takes place in solid solution layers at the junction boundary. Dependence of short circuit current on the illumination intensity L remains linear even if the latter varies by five orders of magnitude. The open circuit voltage changes proportionally to ln$L$ for low light, tending to saturation under high illumination intensity. Using 100 W tungsten lamp as a power source, we were able to measure V_{OC} values at 300 K (see Table 3).

In contrast to the above-discussed integral device characteristics, the spectral data are more variable by depending significantly on heterojunction type and technological conditions during its formation. It is worth noting general features of photosensitivity spectra S, in particular, the fact that S curves are limited by photon energies corresponding to band gap values of heterojucntion components (Fig. 5). The shape of the spectrum and position of its peaks is again defined by the part of the device with major generation of photo-carriers.

Let us analyze this question in detail for a particular case of p-CdTe/n-CdS heterostructure. As one can see from Figure 6, the photosensitivity of a junction synthesized at $T_A = 800 - 1000$ K embraces a wide interval of photon energies and has a blurred peak. This is caused by the presence of the solid solution layer of CdS_xTe_{1-x} responsible for smooth variation of E_g in the area of generation and separation of non-equilibrium carriers.

High-energy edge ends at $h\nu \approx 2.5$ eV, which is close to the E_g of cadmium sulfide. Photons with $h\nu > 2.5$ eV are absorbed in CdS deeper than diffusion length of the minority carriers and thus become lost due to recombination processes. The spectral sensitivity below the band gap of cadmium telluride is caused by non-linearity of band gap dependence for solid solutions CdS_xTe_{1-x} on composition x, which is known [14] to have a minimum at $E_g \approx 1.2$ eV for $x \approx 0.2$; the photons absorbed in this layer will contribute to sensitivity with $h\nu < 1.5$ eV.

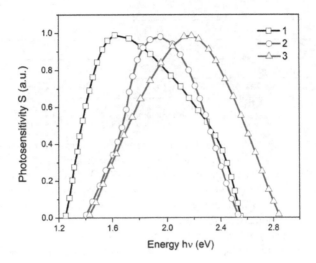

Figure 5. Typical photosensitivity spectra of heterojunctions: 1) p-CdTe/n-CdS, 2) n-CdTe/p-ZnTe and 3) p-CdTe/n-ZnSe at 300 K and illumination from wide-band component side

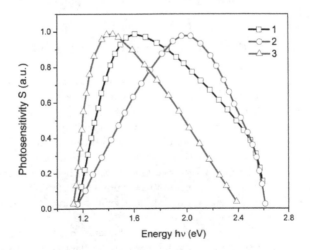

Figure 6. Normalized photosensitivity spectra of p-CdTe/n-CdS heterojunction formed under different temperatures: 1) 800 K, 2) 900 K, and 3) 1000 K

We measured main parameters of heterojunctions related to their possible photovoltaic applications. All measurements were done under AM2 illumination at 300 K and are listed in Table 4. The density of short circuit current was calculated from the experimental I_{SC}, accounting for the effective area of the diodes that is about 0.1 cm^2.

Heterostructure	I_{SC}, mA/cm^2	V_{OC}, V	η, %	FF
p-CdTe/n-CdS	18	0.5	10	0.7
n-CdTe/p-ZnTe	10	0.7	5	0.5
p-CdTe/n-ZnSe	11	0.6	6	0.55

Table 4. Main parameters of solar cells based on heterostructures studied

As one can see from the table, the largest efficiency of 10% corresponds to the photovoltaic device based on p-CdTe/n-CdS, which is below the theoretical limit of 28% due to several reasons. Low fill-factor values FF are caused by a considerable series resistance of dozens of Ohms. This problem can be amended by low-resistance substrates and optimization of ohmic contacts primarily to the thin film layers of cadmium telluride. The benefits of obtained heterojunctions in first place is the low value of thermal efficiency variation $d\eta/dT$ = (0.02-0.03)%/K – that is, four times smaller than that of the silicon solar cells. This is important for solar concentrator applications, when the structure intensively heats. Thus, in spite of modest η, our heterojunctions may compete with GaAlAs ones, because for the same $d\eta/dT$ they are significantly cheaper and simpler in manufacture. It is also worth mentioning that the presence of isovalent impurities in CdTe heterolayers increments radiation stability of the material, which is especially important for open space applications.

4. Heterojunction CdS/PbS cells obtained by chemical bath deposition

Continuing the discussion of solid state solar cells, let us perform optimization of absorber layer keeping the window layer of wide-band CdS, paying a special attention to simplification of deposition technology with an aim to reduce the production costs. Among the main techniques used for the deposition of CdS thin films, we highlight chemical bath deposition (CBD) that produces layers with excellent characteristics because of their compactness and uniformity due to congruent growth and high relative photoconductivity [16, 17]. Furthermore, CBD is a good method for a large area deposition, which is convenient for solar cell fabrication on industrial scale. In CBD technique, the properties of thin films can be controlled by several parameters such as pH of the reaction solution, concentration of the chemical precursors, temperature, deposition time, etc. However, on the other hand, the CBD technique for the deposition of CdS films has serious drawbacks such as large amount of Cd-containing toxic waste produced in the process. Moreover, the classic CBD uses highly volatile and harmful ammonia as the complexing agent in the reaction solution, which can become even more critical in large scale production. These disadvantages catalyze intensive researc aiming to improve the CBD process. For example, ammonia has been substituted with more convenient complexing agents such as ethylendiamine, ethanolamine, triethanolamine, nitrilotriacetic acid and sodium citrate. In particularly, we have developed CBD process based on sodium citrate in place of ammonia [17].

Sodium citrate is a cheap and practically harmless organic compound widely employed in food industry as flavoring or preservative, also as a common ingredient for drinks. We reported main characteristics of CdS films deposited over glass substrates by the partial and complete substitution of ammonia by sodium citrate in the CBD process, resulting in thin films of high crystallinity degree, homogeneity and compactness that performed pretty well as window layers in CdS /CdTe thin film solar cells and as active layers in field effect thin film transistors. Here we would like to discuss CdS window layers obtained by ammonia-free CBD process for CdS/CdTe and CdS/PbS solar cell heterostructures.

The CdS/CdTe and CdS/PbS solar cells were deposited in superstrate geometry onto ITO-coated glass substrates employing two types of chemically-deposited CdS window layers labeled X-CdS and Y-CdS, respectively [18]. The CBD process for Y-CdS films is ammonia-free sodium citrate-based process, consisted in a 100 ml reaction solution prepared in a beaker by the sequential addition of 10 ml of 0.05 M cadmium chloride $CdCl_2$, 20 ml of 0.5 M sodium citrate $Na_3C_6H_5O_7$, 5 ml of 0.3 M potassium hydroxide KOH, 5 ml of a pH 10 borate buffer, 10 ml of 0.5 M thiourea $CS(NH_2)_2$ and deionized water to complete the total volume. The deposition process for X-CdS films consisted in the reaction solution including 25 ml of 0.1 M $CdCl_2$, 20 ml of 1 M $Na_3C_6H_5O_7$, 15 ml of 4 M ammonium hydroxide NH_4OH, 10 ml of 1 M $CS(NH_2)_2$ and deionized water to complete the total volume of 100 ml. In this case, the complexing agent is the mixture of ammonium hydroxide and sodium citrate. In both processes, the beaker with the reaction solution was placed in a thermal water bath at 70 °C. The deposition time was adjusted (20-60 min) to obtain CdS window layers some 100 nm thick. The deposition rates depend on the concentration of the precursors in the reaction solution. It was noticed that the amount of Cd ions is much higher in deposition of X-CdS films – 2.81 mg/ml; for Cd-Y films the numbers are lower – 0.56 mg/ml.

The CdTe thin films on ITO/CdS substrates were deposited by the close-spaced vapor transport-hot wall (CSVT-HW) technique using CdTe powders of 99.99% purity. The deposition of CdTe was performed in Ar/O_2 atmosphere, with each components having partial pressure of 0.05 Torr. The temperatures of the substrate and the source were set to 550 °C and 650 °C, respectively. The deposition process was carried out for 4 minutes. Under these conditions, the resulting thickness of the CdTe layers was approximately 3 μm. After deposition, the CdTe thin films were coated with a 200 nm $CdCl_2$ layer and annealed at 400 °C for 30 min in the air. To create back contact, we deposited by evaporation two layers of Cu and Au with thickness of 20 Å and 350 nm, respectively. The area of the contacts on CdTe side was 0.08 cm^2; after the deposition, the device was annealed at 180 °C in argon atmosphere. The efficiency of CdS/CdTe solar cells was determined from current-voltage measurements under 50 mW/cm^2 illumination.

To produce PbS/CdS solar cells, we deposited PbS film over ITO/CdS substrates by the CBD process that included 2.5 ml of 0.5 M lead acetate, 2.5 ml of 2M NaOH, 3 ml of 1 M thiourea,

2 ml of 1 M triethanolamine and deionized water to bring the total volume to 100 ml. The films were deposited at 70°C for one hour and their thickness was about of 4.2 μm. The solar cell structures were completed with 0.16 cm² printed layer of conducting graphite on the PbS films, serving as back contact. The efficiency of the CdS/PbS solar cells was determined from CVC measurements under 90 mW/cm² illumination.

Figure 7 shows the CVC for the CdS/CdTe solar cells, with X-CdS and Y-CdS window layers. The corresponding performance parameters of both types of solar cells are presented in Table 5. It is observed that the performance of solar cells with X-CdS window layer is better, featuring short circuit current density of 11.9 mA/cm², open circuit voltage of 630 mV, fill factor of 58%, yielding the efficiency of 8.7%.

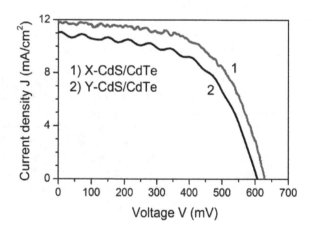

Figure 7. Current density versus voltage measurements under illumination of X-CdS/CdTe and Y-CdS/CdTe solar cells deposited on ITO conductive glass substrates.

Figure 8 presents the CVCs for X-CdS/PbS and Y-CdS/PbS solar cells. The performance parameters determined from these measurements are also given in Table 5. As expected, the performance of these solar cells is much lower because band gap of PbS is smaller, namely 0.4 eV. Nevertheless, the X-CdS window layer performs better also in these solar cells, allowing to reach short circuit current density of 14 mA/cm², open circuit voltage of 290 mV, fill factor of 36% and the efficiency of 1.63%. The low fill factor can be a consequence of high porosity characteristic for semiconductors obtained by the CBD method.

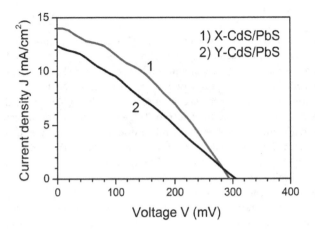

Figure 8. Current-voltage curves for illuminated X-CdS/PbS and Y-CdS/PbS solar cells deposited on ITO conductive glass substrates.

Heterostructure	V_{OC} (mV)	J_{SC} (mA/cm²)	FF (%)	η (%)
X-CdS/CdTe	630	11.9	58	8.7
Y-CdS/CdTe	607	11.1	56	7.5
X-CdS/PbS	290	14	36	1.63
Y-CdS/PbS	310	12.37	28	1.22

Table 5. Parameters of the X-CdS/CdTe and Y-CdS/PbS solar cells with two types of CdS window layers on ITO conductive glass substrates

The analysis of CdS/CdTe and CdS/PbS solar cells given above prove that both X-CdS and Y-CdS are appropriate materials for window layers, with higher efficiency achievable for the solar cells with the X-CdS window. The CBD process for CdS layers required to use both ammonia and sodium citrate as complexing agents. Nevertheless, this variation of CBD is more environmental-friendly with reduced ammonia use due to its partial substitution by sodium citrate. For Y-CdS layers, our optimized CBD process also reduces the amount of cadmium in the reacting solution by the factor of five comparing to the common process of CdS film deposition. Therefore, despite of lower efficiency of cells with Y-CdS window layers, completely ammonia-free process is more convenient for industrial-scale manufacture of CdS/CdTe and CdS/PbS solar cells.

5. Organic solar cells

The "third generation" of photovoltaic technology [19], appearing quite recently, divides into two principal approaches: achieving high efficiencies by creating multiple electron-hole pairs (including thermo-photonic cells) with high cost of the cell or, alternatively, to create very cheap cells with a moderate photovoltaic efficiency (~ 15-20%). Polymer solar cells have a significant impact potential for the second approach. A key point in the development of photovoltaic technology is reduction of cost for large scale production, which stimulates research for alternative materials (such as semiconductors, organics, polymers, heterostructures or composites) for solar cells and photovoltaic devices. The efficiency of inorganic solar cells reaches above 24% due to the use of expensive high purity materials. The production cost can be significantly reduced switching to cheaper constructions including nano-crystalline photoelectrochemical solar cells, pigment sensitivity (dye-sensitized cells), heterojunction polymer/fullerene organic-inorganic hybrid devices and solar cells based on inorganic nanoparticles. These solar cells are the classical example of an electronic device in which organic and inorganic materials complement each other in photovoltaic conversion. The nanomaterials and nanoparticles can also be used for development of energy saving and efficient electronic devices.

The semiconducting conjugated polymers are attractive for their use in photovoltaic cells, since they are strong absorbers and can be deposited onto flexible substrates at a low cost. Cells made from a conductive polymer and two electrodes tend to be inefficient because the photo-generated excitons (mobile excited states) are not separated by the electric field due to differences in the work functions of the electrodes; intensification of such separation helps to improve cell efficiency. Further performance boost can be achieved by optimization of cell design aiming to enhance charge transport and reduce recombination losses.

Polymer photovoltaic devices have a great potential, representing technological alternative to the classical solid-state renewable energy devices. The demand for low cost solar cells catalyzes new approaches and technological developments. In the past years, a significant scientific interest was attracted to solar cells based on organic molecules and conjugated polymers [20], which benefit much from mechanical flexibility and low weight. The polymers' band gap can be easily changed in organic synthesis, allowing production of polymers that absorb light at different wavelengths – which in the case of solid-state photovoltaics was achievable only by creating complicated tandem heterojunctions.

The main operation principles of organic photovoltaic cells differ from those taking place in solid-state semiconductor devices. In the organic material, light absorption leads to generation of excitons; in inorganic cells, illumination produces non-bound electrons and holes. To create photocurrent, it is necessary to separate the exciton into electron and hole before they recombine with each other. In a conjugated polymer, the stabilization of photo-excited electron-hole pairs can be achieved by polymer compounds containing acceptor molecules with electron affinity exceeding that of the polymer, but lower than the corresponding ionization potential. In addition, the highest occupied molecular orbital (HOMO) of the acceptor must have lower energy than those in the conjugated polymer. Under these conditions it becomes energetically favorable for conjugated polymers to transfer photo-excited electrons to the acceptor molecule, keeping the hole at the lowest energy level corresponding to the valence band of the polymer.

Conjugated polymers have de-localized π-electron systems that can absorb sunlight, produce photo-generated charges and offer means for their transport. One of the promising materials from the family of conjugated polymers used for solar cell applications is the poly(3-hexylthiophene) P3HT [21, 22] with side chains that make it soluble in common organic solvents (Fig. 9), which allows material deposition by wet processing techniques such as spin coating (rotational coating), dip coating [23], ink jet printing [24, 25], screen printing and micromolding [26, 27]. All these methods can be performed at room temperature, normal atmospheric pressure and can be applied to the flexible substrates [28], paving an attractive route for mass-scale production of large-area solar cells at low cost.

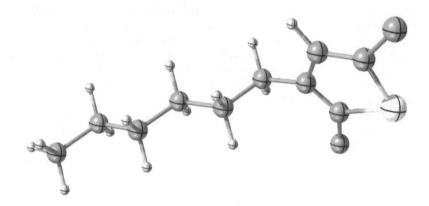

Figure 9. Chemical structure of poly (3-hexylthiophene) P3HT.

The optical bandgap of P3HT is about 1.9 eV that approaches the spectral peak of 1.8 eV (700 nm) of solar light corresponding to terrestrial illumination conditions of AM1.5. P3HT also has high absorption coefficient permitting efficient processing of light with wavelengths up to 650 nm using a film that is only 200 nm thick. The photoactive layer is composed by a heterojunction of two organic semiconductors. Illumination generates excitons that become separated at the junction, producing carrier flow that is collected at the contacts (Fig. 10).

One of the ways to increase current output of the cell is to improve light absorption in the photoactive layer, which can be achieved by reducing the band gap of the polymer. The conjugated polymers, characterized by a high value of absorption coefficient (10^5 cm^{-1}) look as promising candidates in this regard. While the crystalline silicon cells should be approximately 100 µm thick for efficient absorption of the incident light, organic semiconductors with direct bandgap will have the similar performance with reduced thickness of 100-500 nm. However, conjugated polymers are usually characterized with large gap values that are not always sufficient for efficient absorption.

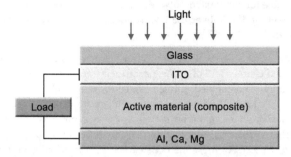

Figure 10. Photovoltaic solar cell: organic material sandwiched between two electrodes. The electrons are collected at the rear metal contact and the holes – at the front ITO contact.

In most organic semiconductors, excitons are comparatively tightly bound and do not dissociate easily. That is why it is useful to create a heterojunction of materials with distinct electron affinities and ionization energies to favor exciton dissociation. In this way, the electron is accepted by the material with higher affinity and the holes proceed to the material with lower ionization energy, producing the effect of a local field separating carriers at the junction. When the donor molecule is excited, an electron is transferred from HOMO to lowest unoccupied molecular orbital (LUMO), forming a hole. If electron-hole pair recombines, luminescence is produced. However, if the LUMO of the acceptor is small enough compared to that of the donor, the excited electron will end up at acceptor's LUMO and the carriers originating from dissociated exciton will be separated (Figure 11).

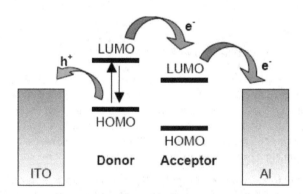

Figure 11. Exciton dissociation at the donor-acceptor interface

The heterojunction cells will be efficient in dissociation of excitons at the interface area [29], requiring generation of an exciton within its diffusion length from the interface. As diffusion length value is about 10 nm, it limits the thickness of light-absorbing layer. On the other hand, for the majority of organic semiconductors it is necessary to have a film of more than

100 nm thick to ensure sufficient light absorption, which, in turn, lowers the number of excitons that can reach the interface. For this reason, it is proposed to obtain dispersed (bulk) heterojunctions, schematically depicted in Fig 12.

One of the main problems to create such type of solar cells concerns miscibility of the components. Conjugated systems – including polymer conjugates and dyes – are usually immiscible, so that even a completed solar cell may represent a non-equilibrium system. To improve the situation somehow, it is proposed to use the spin coating technique that is characterized with rapid solvent evaporation, requires little parameter adjustments and tolerates a wide range of viscosities. The deposition process usually has three steps: dispersion of nanoparticles in a solvent, mixing them with a polymer and finally molding of the compound. Other techniques, such as co-evaporation and co-sputtering allow better control over morphology of the material; however, they are quite costly due to particular temperature and pressure requirements (e.g., deposition should be carried out in vacuum).

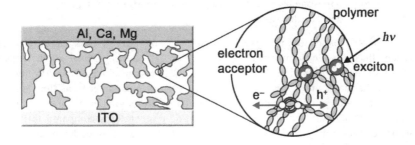

Figure 12. Architecture of a photovoltaic cell with a dispersed heterojunction formed by conjugated polymer with transparent ITO front electrode and Al, Ca, Mg rear electrode.

A new generation of solar cells called hybrid polymer solar cells attracts a considerable interest [30-31]. Recent studies have shown that nanoparticles incorporated into photoactive layer improve light absorption and increase photocurrent. Nanoparticle polymer-based photovoltaic cells have a long term potential for decreasing the cost and improve device efficiency. The maximum efficiency reported to date is around 5.55% (with theoretical predictions of about 10%). Incorporation of 5-10 nm nanoparticles of gold into poly (9.9-dioctylfluorene) results in significant improvement of cell efficiency and oxidation stability.

6. Heating issues and their treatment

Direct absorption of energy from the sunlight faces another considerable problem connected with operation of electronic devices under elevated temperatures due to variation of band gap of semiconductors, increase of thermal noise, etc. A partial solution to this problem can be offered by the use of passive/active radiators that will dissipate some of excessive heat. Further development of this idea leads to the use of hybrid systems that allow to process part of solar light via photoelectric conversion, storing thermal energy, for example, by heating water that can be used for household needs. This, however, does not allow to reduce the problems of high-temperature operation of an electronic device (solar cell) considerably. A useful solution can be glimpsed from the nanoelectronics, which also have problems with removal of excess heat from the integrated circuits. Thermal management in these devices has become problematic because faster and denser circuits are required to meet the modern needs, which, in turn, produces even more heat. Localized areas of high heat flux influences the performance at both the chip and the board levels for the current nanotechnologies.

Key concepts like waste heat recycling or waste heat recovery are the basic ideas in the design of the newest heat protection and dissipation systems. The potential applications of the thermoelectric devices are thus enormous. Thermoelectricity is the revolutionary technology that is currently under intense development aiming to find a solution to thermal management problem and protection of small-scale systems. However, due to relatively low efficiency (around 10%), thermoelectric cooling is generally only used in small systems; the new concepts based on nanoscale heat transfer bring a new opportunity to widen the application horizons for thermoelectric devices.

As expected, technology scaling significantly impacts power dissipation issues. The scale-connected effects for silicon-on-insulator (SOI) technology affect electrical properties of the material. Joule heat generated in SOI transistors may compromise long term reliability of the device. The thermal conductivity of the channel region of nanometer transistors is significantly reduced by phonon confinement and boundary scattering. This increases the thermal resistance of the device, leading to higher operating temperatures compared to the bulk transistors of the same power input. However, ballistic transport between the material boundaries impedes device cooling, so that temperature-dependent parameters of the device such as source-drain current and threshold-current will increase significantly, generating much Joule heat that will eventually lead to accelerated temperature degradation of the gate dielectric [32, 33]. The recently developed germanium-on-insulator (GeOI) technology combines high carrier mobility with the advantages of the SOI structure, offering an attractive integration platform for the future CMOS devices [34].

SiGe nanostructures are very promising for thermoelectric cooling of microelectronic devices and high-temperature thermoelectric power generation. It has been demonstrated that the thermal conductivity is significantly reduced in super-lattices [35-37] and quantum dot super-lattices [38-40]. A self-organized set of vertically stacked Si/Ge quantum dots is a good alternative to induce artificial scattering of phonons and reduce the thermal conductivity. One of the ways to increase scattering even more involves creation of the structure where with uncorrelated vertical positions of quantum dots, reducing the effective thermal path of the phonon within the Si layer as shown in Figure 13. The phonons travelling by the pathway laid by Ge quantum dots will experience higher scattering than phonons travelling through the Si pathway only. This effect becomes even more efficient because, in fact, the phonons spreading through Ge quantum dots will suffer a sequence of scattering events from one dot to another. As one can see from the figure, temperature-dependent cross-plane thermal conductivity reduces dramatically in Ge quantum dot superlattices depending on vertical correlation between dots, with at least twice lower thermal conductivity value obtained for the case of uncorrelated dot structures in comparison with well-aligned dot array with the same vertical spacing of 20 nm. The same result can be confirmed by Raman spectroscopy [41].

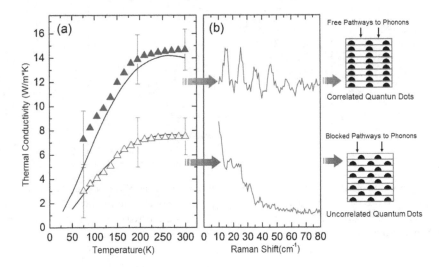

Figure 13. Dependence of thermal conductivity on the temperature for correlated and uncorrelated Ge quantum dot structures

Thin film single-crystalline GeOI structures may be considered as potential candidates in the field of CMOS microelectronics improving thermal performance of transistors due to superior mobility of carriers compared to other semiconductors. Recent predictions in the thermal conductivity of ultra-thin germanium films suggest that the small bulk mean free path of Ge will induce a weaker effect on the boundary scattering [42]. This thermal behavior is an additional reason that makes GeOI structures competitive with SOI for the case of small-size thin film devices at the cutting edge of the technology. For example, it has been reported from Monte Carlo modeling that electro-thermally optimized GeOI structures should be 30% more productive than the best SOI device examples [34]. In addition, the higher mobility of germanium implies that GeOI devices might support the same amount of current at lower operating voltage, so that the dissipated power is expected to be lower [42].

Figure 14 shows a plot of the intrinsic out-of-plane thermal conductivity variation with thickness at 300 K. In spite of low thermal conductivity of bulk Ge compared to that of Si ($\kappa_{Ge}/\kappa_{Si} \sim 0.4$), ultra-thin films of germanium has smaller thermal conductivity due to reduced mean free path. Hence, ultra-thin films of Ge suffer from a lower reduction of the thermal conductivity compared to ultra-thin films of Si, which makes germanium-on-insulator structures promising candidates for devices with reduced self-heating effects compared to silicon-on-insulator structures [43].

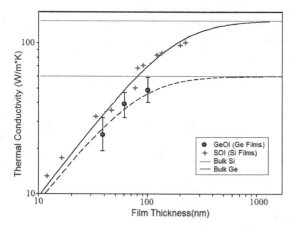

Figure 14. Dependence of thermal conductivity κ on film thickness for different materials.

The experience obtained with aforementioned thermoelectric applications can be successful-ly applied to the field of photovoltaics, forming efficient heat-draining layers either on the frontal surface of solar cells that suffer highest temperature increase or over the contact grid that, in addition to solar heating, also experiments Joule heat.

7. Conclusions

This chapter addresses a wide number of topics concerning thin-film solar cells. It was shown that the numerical modeling of current transport in AlGaAs/GaAs heterojunctions al-lows to determine optimal aluminum content (that defines band gap difference of the junc-tion components) ensuring the most efficient processing of the incident light flux by window and absorber layers. We found that for very thin window layers the proximity of the junction area to the surface of the cell has more prominent role, allowing the embedded field of space charge region to function more intensively in separation of photo-generated currents and reducing the effect of recombination phenomena.

The question concerning quality improvement of junction boundary is reflected in the sec-ond section that presents results concerning the use of isovalent substitution method for manufacturing of heterojunction solar cells. As substituted layers grow into the substrate, we obtain a smooth transition of one material into another that reduces the difference of lat-tice constants and thermal expansion coefficients, both of which are of high importance for photovoltaic devices. Good efficiency values for non-optimized cells without any special an-ti-reflection coatings and with considerable series resistance paves the way for future im-provements.

A special attention is being paid to creation of cheap and environment-friendly technologies for solar cells; this point is illustrated with an example of CdS/PbS heterojunctions created by ammonia-free chemical bath deposition. It is thought that these results will be interesting for large-scale industrial production of solar cells.

We address the important questions of organic solar cells, which nowadays attract much attention of the scientific community. These photovoltaic devices has lower efficiency in comparison with silicon or tandem multi-junction cells, but they are incomparably cheaper and can use flexible substrates, which opens completely new and wide horizons for their possible applications. We also discuss the problems of the proper choice of organic material for the active element of the cell.

Finally, solar cells are always overheated due to exposure to a direct sunlight, which makes a considerable problem concerning degradation of device parameters under prolonged operation under elevated temperatures, as well as mechanical stability of the cell due to thermal expansion of its components. We propose to make some useful parallels with nanoelectronics, which recently received promising solutions in a form of thermoelectric heat transfer managing devices. We hope that the similar techniques could be applied to solar cells, offering good results with temperature control for photovoltaics, especially those operating under concentrated sunlight conditions.

Author details

P.P. Horley[1], L. Licea Jiménez[1,2], S.A. Pérez García[1,2], J. Álvarez Quintana[1,2], Yu.V. Vorobiev[3], R. Ramírez Bon[3], V.P. Makhniy[4] and J. González Hernández[1]

1 Centro de Investigación en Materiales Avanzados, Chihuahua - Monterrey, México

2 GENES Group of Embedded Nanomaterials for Energy Scavenging, Apodaca, México

3 Centro de Investigación y Estudios Avanzados Unidad Querétaro, Querétaro, México

4 Yuri Fedkovych Chernivtsi National University, Chernivtsi, Ukraine

References

[1] Green MA. Silicon Solar Cells: Advanced Principles and Practice. Sydney: Bridge; 1995.

[2] Gonçalves da Silva C. The Fossil Energy/Climate Change Crunch: Can we Pin our Hopes on New Energy Technologies?, Energy 2010; 35 1312–1316.

[3] Green MA, Emery K, Hishikawa Y, Warta W, Dunlop ED. Solar Cells Efficiency Tables (Version 38), Progress in Photovoltaics: Research and Applications 2011; 19 565-572.

[4] Hillhouse HW, Beard MC. Solar Cells from Colloidal Nanocrystals: Fundamentals, Materials, Devices, and Economics. Current Opinion in Colloid & Interface Science 2009; 14 245–259.

[5] Chopra K, Das A. Thin Film Solar Cells. New York: Plenum Press, 1986.

[6] Farhenbruch A, Bube R. Fundamentals of Solar cells. New York: Academic Press, 1983.

[7] Pauwels HJ, De Visschere P, Reussens P. Analysis of Generation in Space Charge Regions of Solar Cells. Solid State Electronics 1978; 21 775-779.

[8] Serdyuk VV. Physics of Solar Elements. Odessa: Logos, 1994.

[9] Hovel HJ, Woodwall JM. $Ga_{1-x}Al_xAs$-GaAs p-n Heterojunction Solar Cells. Journal of Electrochemical Society 1973; 120 1246.

[10] Hulstorm R, Bird R, Riodan C. Spectral Solar Irradiance Data Sets for Selected Terrestrial Conditions. Solar Cells 1985; 15 365.

[11] Ponpon JP. A Review of Ohmic and Rectifying Contacts on Cadmium Telluride. Solid State Electronics 1985; 28(7) 689-706.

[12] Sharma BL, Purohit RK. Semiconductor Heterojunctions. NY: Pergamon Press, 1974.

[13] Makhiny VP, Baranyuk VE, Demich NV, et al. Isovalent Substitution: a Perspective Method of Producing Heterojunction Optoelectronic Devices. Proc. SPIE 2000; 4425 272.

[14] Ryzhikov VD. Scintillation Crystals of A^2B^6 Semiconductors. Moscow: NIITEHIM, 1989.

[15] Nakada T, Mitzutani M, Hagiwara Y, Kunioka A. High-efficiency Cu(In,Ga)Se2 Thin Film Solar Cells with a CBD-ZnS Buffer Layer, Solar Energy Materials and Solar Cells 2001; 67 255–260.

[16] De Melo O, Hernandez L, Zelaya-Angel O, Lozada-Morales R, Becerril M, Vasco E. Low Resistivity Cubic Phase CdS Films by Chemical Bath Deposition Technique, Applied Physics Letters 1994; 65 1278–1281.

[17] Ortuño-López MB, Valenzula-Jauregui JJ, Sotelo-Lerma M, Mendoza-Galvan A, Ramírez-Bon R. Highly Oriented CdS Films Deposited by an Ammonia-Free Chemical Bath Method. Thin Solid Films 2003; 429 34–39.

[18] Hernández-Borja J, Vorobiev YV, Ramírez-Bon R. Thin Film Solar Cells of CdS/PbS Chemically Deposited by an Ammonia-Free Process. Solar Energy Materials & Solar Cells 2011; 95 1882–1888.

[19] Green MA. Third Generation Photovoltaics: Ultra-High Efficiency at Low Cost. Berlin: Springer-Verlag, 2003.

[20] Krebs FC. Fabrication and Processing of Polymer Solar Cells: a Review of Printing and Coating Techniques. Solar Energy Materials and Solar Cells 2009; 93 394–412.

[21] Zhou QM, Hou Q, Zheng L et al. Fluorene-based Low Band-gap Copolymers for High Performance Photovoltaic Devices. Applied Physics Letters 2004; 84 1653–1655.

[22] Xin H., Ren G., Sunjoo Kim F., Jenekhe SA. Bulk Heterojunction Solar Cells from Poly(3-butylthiophene)/ Fullerene Blends. Chemical Materials 2008; 20 6199–6207.

[23] Wang G, Swensen J, Moses D, Heeger AJ. Increased Mobility from Regioregular Poly(3-hexylthophene) Field-Effect Transistors. Journal of Applied Physics 2003; 93(10) 6137.

[24] Hebner TR, Wu CC, Marcy D, Lu MH, Sturm JC. Ink-jet Printing of Doped Polymers for Organic Light Emitting Devices. Applied Physics Letters 1998; 72(5) 519.

[25] Chang SC, Liu J, Bharathan J, Yang Y, Onohara J, Kido J. Mutlicolor Organic Light-Emitting Diodes Processed by Hybrid Inkjet Printing. Advanced Materials 1999; 11 734.

[26] Pschenitzka F, Sturm JC. Three-color Organic Light-emitting Diodes Patterned by Masked Dye Diffusion. Applied Physics Letters 1999; 74(13) 1913.

[27] Rogers JA, Bao Z, Raju VR. Nonphotolithographic Fabrication of Organic Transistors with Micron Feature Sizes. Applied Physics Letters 1998; 72(21) 2716.

[28] Gustafsson G, Cao Y, Treacy GM, Klavetter F, Colaneri N, Heeger AJ. Flexible Light-Emitting Diodes Made from Soluble Conducting Polymers. Nature 1992; 357 477-478.

[29] Saunders BR, Turner ML. Nanoparticle–Polymer Photovoltaic Cells. Advances in Colloid and Interface Science 2008; 138 1–23.

[30] Sariciftci NS, Smilowitz L, Heeger AJ, Wudl F. Photoinduced Electron Transfer from a Conducting Polymer to Buckminsterfullerene. Science 1992; 258 1474-1476.

[31] Yu G, Gao J, Hummelen JC, Wudl F, Heeger AJ. Polymer Photovoltaic Cells: Enhanced Efficiencies via a Network of Internal Donor-Acceptor Heterojunctions, Science 1995; 270 1789-1791.

[32] Fiegna C, Yang Y, Sangiorgi E, O'Neill AG. Analysis of Self-Heating Effects in Ultra-thin-Body SOI MOSFETs by Device Simulation. IEEE Transactions on Electron Devices 2008; 55(1) 233-244.

[33] Pop E, Sinha S, Goodson KE. Heat Generation and Transport in Nanometer-Scale Transistors. Proccedings of the IEEE 2006; 94 (8) 1587-1601.

[34] Lee ML, Fitgerald EA. Strained Si/Strained Ge Dual-Channel Heterostructures on Relaxed $Si_{0.5}Ge_{0.5}$ for Symmetric Mobility p-type and n-type Metal-Oxide-Semiconductor Field-Effect Transistors. Applied Physics Letters 2003; 83(20) 4202.

[35] Lee SM, Cahill DG, Venkatasubramanian R. Thermal Conductivity of Si-Ge Superlattices. Applied Physics Letters 1997; 70 2957.

[36] Lu X, Chu J. Lattice Thermal Conductivity in a Si/Ge/Si Heterostructure. Journal of Applied Physics 2007; 101 114323.

[37] Kiselev AA, Kim KW, Stroscio MA. Thermal Conductivity of Si/Ge Superlattices: A Realistic Model with a Diatomic Unit Cell. Physical Review B 2000; 62 6896.

[38] Liu JL, Khitun A, Wang KL, et al. Cross-Plane Thermal Conductivity of Self-Assembled Ge Quantum Dot Superlattices. Physical Review B 2003; 67 165333.

[39] Shamsa M, Liu W, Balandin A, Liu J. Phonon-Hopping Thermal Conduction in Quantum Dot Superlattices. Applied Physics Letters 2005; 87 202105.

[40] Lee LL, Venkatasubramanian R. Effect of Nanodot Areal Density and Period on Thermal Conductivity in SiGe/Si Nanodot Superlattices. Appl. Phys. Lett. 2008; 92 053112.

[41] Alvarez-Quintana J, Alvarez X, Rodriguez-Viejo J, Jou D, Lacharmoise PD, Bernardi A, Goñi AR, Alonso MI. Cross-plane Thermal Conductivity Reduction of Vertically Uncorrelated Ge/Si Quantum Dot Superlattices. Applied Physics Letters 2008; 93 013112.

[42] Pop E, Chu CO, Dutton R, Sinha S, Goodson K. Electro-Thermal Comparison and Performance Optimization of Thin-Body SOI and GOI MOSFETs. IEDM Technical Digest. IEEE International 2004; 411-414.

[43] Alvarez-Quintana J, Rodríguez-Viejo J, Alvarez FX, Jou D. Thermal Conductivity of Thin Signle-Crystalline Germanium-on-Insulator Structures. International Journal of Heat and Mass Transfer 2011; 54 1959–1962.

Conceptual Study of a Thermal Storage Module for Solar Power Plants with Parabolic Trough Concentrators

Valentina A. Salomoni, Carmelo E. Majorana,
Giuseppe M. Giannuzzi, Rosa Di Maggio,
Fabrizio Girardi, Domenico Mele and
Marco Lucentini

Additional information is available at the end of the chapter

1. Introduction

Technologies and methods for thermal energy storage have been well tested in CSP - Concentrated Solar Power – plants [1, 2]. Solar tower plants (e.g. Solar Two, USA) and advanced parabolic trough plants (e.g. Archimede by ENEA, Italy) use molten salts both as heat transfer and thermal storage fluid. Differently, traditional trough plants (e.g. Andasol, Spain) distinguish the fluid through the solar field (synthetic oil) from the one used in the storage system (molten salt). Hence, storage applications have only been proven in liquid state and in large scale plants.

Concrete is the generally preferred "solid" material for its low cost and good thermal conductivity, already tested at the Platform Solar of Almeria (Spain) and by DLR (Germany) revealing an appropriate response to the specific use, among which a structural stability [3, 4].

Thermal storage of sensible heat using concrete is at present a known procedure, but applications are still limited and some variables (e.g. concrete durability, concrete mixing, etc.) are unclear or not appropriately defined. Briefly, limitations of existing solid thermal energy storage systems include: existing systems are conceived for operating in big capacity plants (skilled personnel and technical infrastructure available); concrete mixtures: actual research is focused on thermal performances optimization regardless costs and durability issues; hygro-thermal issues, not well defined, has not been treated and disseminated.

Nowadays, new mixtures are under study but their characterization seems to be not completed.

Some attempts have been tried incorporating components to increase the conductivity of the composite materials as graphite but this solution increases the cost, however no experience exists in concrete. The technology of concrete production can also contribute to the optimisation of the type of concrete for TES (Thermal Energy Storage), as for instance self compacted concrete.

The main technical objectives of the authors' current research include: 1) development of an appropriate concrete mixing, optimizing chemical-physical and durability performances in a temperature range up to 300°C; 2) thermal sizing of the storage module and its integration within a CSP system. Such results will be obtained through various activities, for some part reported in this Chapter:

- verification of the (hydro)-thermo-mechanical response for the storage module, in its start-up and exercise stages;

- realization and experimentation of a prototype at reduced scale;

- conceptual project of a thermal storage module, including a synthesis of both technical and economical aspects.

Among the innovative research and development items, we recall:

a. The system is constituted by modular blocks made by innovative concrete material whose mix design is being investigated [5]. The constitutive characterization of the concrete mixture chosen as thermal storage, once passed theoretical to experimental investigation, will represent a fundamental cognitive relapse in the context of cementitious materials subjected to high temperature [6]. In its basic aspects it will be inspired by complementary applications within an already operative project to which ENEA, and the Universities of Padua, Trento and Rome, Italy, are contributing. The research outcomes have already proved the possibility to satisfy the features and behaviour requested for such an application in terms of thermal and mechanical characteristics. It is anyway necessary to deepen the structural behaviour and aspects regarding durability and possible local small damaged spots which can bring to a possible loss in functionality of the component, being nowadays an open question;

b. The tubes in which the heat-transfer fluid flows are immersed inside the concrete matrix. The up-to-now adopted constructive solution used by DLR consists in a tube bundle with adequately supported pipe fittings and in a particular distribution within concrete casting [3, 4, 7, 8]. A sort of concrete mono-block comes out: the free water contained within its concrete matrix hardly leaks, especially during its first heating phase. This can bring to high pressure gradients, causing complete component damage or reductions in thermal performance.

Our research activity will match a wide application range of expertise in order to avoid original problems of module cracking. A balanced new-type concrete mixture, derived from previous researches, will be cast under innovative procedures. Piping would derive from heat exchanger applications with specials interface layers also experienced in nuclear plants. Start-up and operational phases would be conceived to limit matrix pressure stress.

2. The SOLTECA Project

The SOLTECA Project contextualizes TES in solid media within the Italian context; the main objectives of the Project include:

- development of an appropriate concrete mixing so to optimize its chemical-physical properties, durability and performance at temperatures between 80-300°C;

- thermal design of a storage module and its integration in CSP systems.

The Project is addressed to solar plants with innovative aspects and characteristics such as: small size plants with a peak power between 0.5-5 MWe, to be more easily placed in the territory; sensible heat storage in concrete with ad-hoc mixing; modular structure of the storage system with elements of reduced dimensions so to be pre-cast and to limit the efforts during degas phases at start-up; capability of returning energy to the heat-transfer fluid at temperatures between 120-300°C; limitation of the storage specific cost between 20-30 €/kWh_th; use of a heat-transfer fluid with low environmental impact (possibly water); coupling of the solar plant and the storage system with ORC (Organic Rankine Cycles) groups and biomass plants.

It is to be underlined that the maximum operational temperature for concrete is fixed at about 300°C; if water is chosen as heat-transfer fluid, the maximum temperature will be appropriately reduced to limit the service pressure.

As stated, TES is based on CSP systems and the technology has been developed for power plants (SPP, Solar Power Plants) of big dimensions; most of the solar source power comes from the so-called "sun-belt", i.e. the most irradiated area of the planet such as North Africa and Middle East. Anyway the concentrated solar technology can be adopted even in Southern Europe and Italy by integrating it with other renewable technologies which will have to contribute to the growing European demand for "green electricity".

The thermodynamic solar plants produce energy following the same process as for the conventional vapour plants, but using solar radiation as a primary energetic source in agreement with the scheme of Figure 1 for a sensible heat storage system in solid media where energy charge and discharge occur via a heat-transfer fluid circulating in an embedded piping system.

The prior objective in the design phase of a SPP is the definition of the plant configuration and of the single component dimension so to minimize the cost of the produced energy unit, i.e. the Levelized Energy Cost (LEC).

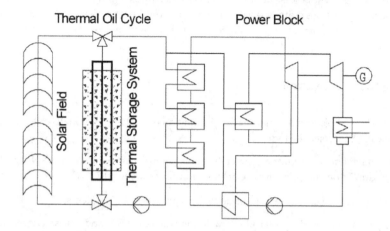

Figure 1. General scheme of a thermodynamic solar plant [8].

3. Storage issues for renewable energies

One of the major restraint to the development of energetic systems from renewable sources is given by the uncertainty with which these sources give energy to the plant. Hence the availability of storage systems is a key-factor in both development and success of a technology for the production of heat or electric energy by using solar radiation for the daily and the long-term use. Nowadays systems able to store high quantities of energy with reduced costs and sizes compatible to those of the serviced plants are not defined yet; the basic principle according to which a storage energy system is used is that of trying to align the generation curve of these plants with the one coming from the users requirements. Whereas the former is linked to the -seasonal or daily- variation of solar energy, the latter depends on the users type to be served (industrial, domestic or services) so that there is no connection between the two.

In general, the main functions performed by a storage system within a SPP are:

a.   buffering during variable sunshine periods,

b.   time shifting in using available radiation,

c.   increment in the annual Capacity Factor and consequent reduction in the cost of produced energy,

d.   more regular production of energy.

Storage allows for facing the oscillatory trend of incoming radiation (Figure 2) and contemporaneously shifts the use of radiation, in excess during the central hours of the day, towards the late-evening hours. A storage capacity able to allow the plant for operating at its nominal power for 24 hours is theoretically but not practically reachable; following the minimum LEC criterion, the storage and the collectors system sizes must be simultaneously correlated and optimized. As shown in Figure 3, a variety of energy storage types is possible, some already widely used (e.g. in the hydroelectric field), others in development.

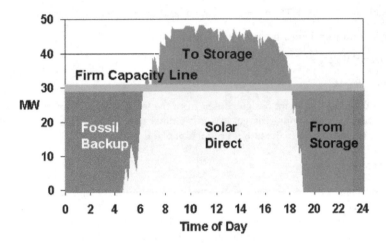

Figure 2. Daily operation for a solar plant with storage system and integration [3].

A complete storage process involves at least three phases: charge, conservation of the stored energy and discharge. In real systems, some of the described steps can occur simultaneously and each of them can even occur more than once for each storage cycle, so that modelling these components becomes a very complicated matter.

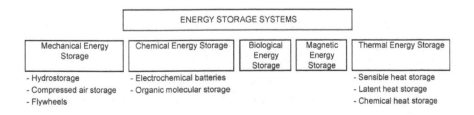

Figure 3. Types of storage system [9].

Multiple aspects must be considered during the design phase for thermal storage systems; from the technical viewpoint, the most relevant specifications are:

- high energy density in the storage medium (storage capacity);

- good heat transfer between Heat Transfer Fluid (HTF) and storage medium (efficiency);

- chemical and mechanical stability of the storage medium (a high number of charge and discharge cycles must be sustained);

- compatibility between HTF and heat exchanger and/or storage material (safety);

- full reversibility of charge/discharge cycles (durability);

- low thermal leakages;

- control easiness.

Moreover, the most important design criteria from the technological point of view are: operational strategies, maximum charge, loss in nominal temperature and specific enthalpy during charge, integration in the power production plant.

3.1. Types of heat storage

Sensible heat storage: thermal energy can be stored via the induced temperature variation in the material, corresponding to an internal energy variation for the sensible heat storage. Among solid materials, e.g. concrete is chosen in the various applications for its low cost, availability and easy workability; additionally, it is a material with high specific heat, good mechanical properties (e.g. compressive strength), thermal expansion coefficient close to the steel one (material used for piping) and high mechanical strength to cyclic thermal loads.

Latent heat storage: thermal energy for some substances can be stored in a nearly-isotherm manner as latent heat linked to phase changes; the materials used for such a technology are named Phase Change Materials (PCM). PCM allow for accumulating high quantity of energy in relatively small volumes, so resulting among the solid media at lowest cost for the various storage concepts.

3.2. Materials for sensible heat storage

The cumulated thermal energy within certain mass of material can be notoriously expressed as

$$Q = \rho \cdot \bar{c}_p \cdot V \cdot \Delta T \tag{1}$$

where it is just to be observed that the specific heat has a mean value within the exercise temperature range.

As previously reported, for being a material usable in a TES-type application, a low cost is needed as well as a good thermal capacity $\rho \bullet c_p$. An additional fundamental parameter for sensible heat TES is the velocity at which heat can be released or extracted; such a character-istic is function of the thermal diffusivity

$$\lambda = \frac{k}{\rho \cdot \overline{c}_p} \tag{2}$$

in which k is the material thermal conductivity.

Concrete and ceramic materials, both tested at the Platforma Solar de Almeria (PSA), present appropriate characteristics for being adopted as sensible heat storage media.

Table 1 shows the main properties for the most frequent solid storage materials in literature; to improve the "soft" characteristics of an ordinary concrete, a high temperature concrete (HTC) has been studied whose features have been compared to those of a castable ceramic material (Table 2).

Storage medium	Temperature		Average density (kg/m³)	Average heat conductivity (W/m K)	Average heat capacity (kJ/kg K)	Volume specific heat capacity (kWhₜ/m³)	Media costs per kg (US$/kWhₜ)	Media costs per kWhₜ (US$/kWhₜ)
	Cold (°C)	Hot (°C)						
Sand-rock-mineral oil	200	300	1700	1.0	1.30	60	0.15	4.2
Reinforced concrete	200	400	2200	1.5	0.85	100	0.05	1.0
NaCl (solid)	200	500	2160	7.0	0.85	150	0.15	1.5
Cast iron	200	400	7200	37.0	0.56	160	1.00	32.0
Cast steel	200	700	7800	40.0	0.60	450	5.00	60.0
Silica fire bricks	200	700	1820	1.5	1.00	150	1.00	7.0
Magnesia fire bricks	200	1200	3000	5.0	1.15	600	2.00	6.0

Table 1. Properties of solid storage media [9].

Material	Castable ceramic	High temperature concrete
Density [kg/m³]	3500	2750
Specific heat at 350 °C [J/kg K]	866	916
Thermal conductivity at 350 °C [W/m K]	1.35	1.0
Coefficient of thermal expansion at 350 °C [10^{-6}/K]	11.8	9.3

Table 2. Comparison between castable ceramic and HTC

4. Concrete storage

The German Aerospace Centre (DLR) within the Project "Midterm Storage Concepts—Fur-ther Development of Solid Media Storage Systems" developed from 2001 to 2003 has al-ready started the development of sensible heat storage-based technology in solid media. The main points of the Project were the development of an efficient material for heat storage and the technology experimentation with a test unit with the size of 350 kWh [3, 4, 7, 8].

Hence such a technology is known and studied but it is not contextualized to other scenarios, as well as uncertainties exist in term of concrete mixing and durability.

Solid storage systems adopting concrete fall in the category of passive storage systems which are generally "double-material": HTF flows in the storage just to charge and discharge a solid material. HTF transfers the heat received from the source of primary energy (the sun) towards the storage module during the charge process and receives energy from the module itself during discharge according to the scheme of Figure 4; such system are even named *regenerators*. The principal disadvantage of regenerators is that the fluid temperature decreases during discharge as the storage medium chills. An additional problem is that heat transfer is relatively low.

The storage medium contains a piping exchanger for the heat transfer from HTF to the storage and *viceversa*, as depicted in Figure 4. Such a heat exchanger represents a significant part of the costs for the whole investment of the plant. The definition of the geometric parameters, as diameter and number of pipes, is also fundamental for the final result from the performance viewpoint.

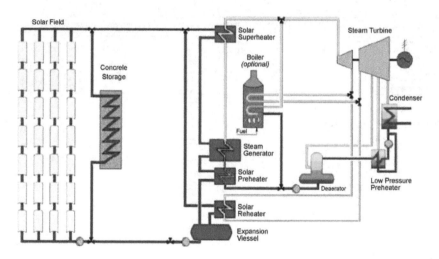

Figure 4. Integration scheme of an indirect storage through solid medium [3, 9].

The advantages of concrete storage systems are:

- low costs of storage media,

- high thermal exchange fluxes within the solid (or from this one towards outside) due to the contact between piping and concrete,

- easy workability of the material,

- low degradation of heat transfer between exchanger and storage medium.

Among the main disadvantages:

- costs increment for exchanger and engineering in general,

- long term instability,

- effective reachable charge level.

With reference to Figure 5, temperature profiles of the material are reported along the channel axis; it is to be noticed that areas underlying the curves are proportional to the quantity of energy if we refer to Equation (1). According to this expression the component's storable energy is proportional to the existing ΔT between the condition of complete charge and discharge. If all the material were first heated and successively uniformly cooled, its storage capacity would be proportional to the area limited by the two horizontal dashed lines. Considering that, in real transients within material, thermal fields develop with the shown qualitatively profiles, a reduced capacity of the system comes out, i.e. there are zones of material which do not contribute to the process (thermally inactive). Hence the objective of the design is to reduce these zones and the parameters on which it is possible to intervene are material itself (via the diffusivity) and geometry (distance between pipes and exchange coefficient).

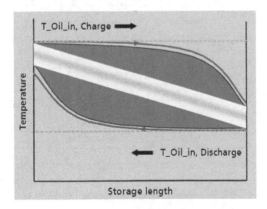

Figure 5. Storable energy within the solid [8].

5. Pre-design procedure

5.1. Definition of storage operational conditions

As stated, even if concrete TES systems have been already studied, they appear to be still in a development phase, so that preliminary analyses are necessary to define design procedures and guidelines for future upgrades.

A simplified approach is here proposed for modelling concrete based on a continuum material model with constant parameters. Considering the characteristics of the plant, an initial design of the different components is performed so to define the operational storage conditions.

With reference to a fixed design plant point, i.e. by assuming in input an effective radiation value and fixing a Solar Multiple value (ratio between the thermal power delivered by the collectors in the design conditions and the nominal power requested by the user),

$$SM = \frac{P_{solar\_field}}{P_{power\_block}}\Bigg|_{Design\_point} \tag{3}$$

the preliminary analysis leads to the parameters of Table 3.

PLANT CHARACTERISTICS		
Nominal radiation	[W/m²]	700
Solar multiple		2
Plant thermal power	[kW]	4500
m_{nomTES}	[kg/s]	11.33
V_{fluid}	[m/s]	1
Heat-transfer fluid		H_2O
Charge time (t_c)	[h]	1
CONCRETE PARAMETERS		
ρ	[kg/m³]	2666
c_p	[J/kg °C]	800
k	[W/m °C]	2
STORAGE EXTREME TEMPERATURES		
T_{in_charge}	[°C]	175
T_{out_charge}	[°C]	125
$T_{in\_discharge}$	[°C]	80
$T_{out\_discharge}$	[°C]	130
Complete charge level		90% ΔT

Table 3. Design characteristics and parameters.

For the material a conductivity value higher than the real one has been assumed, i.e. 2 W/m °C, so to directly refer to a value of k close to the fixed target one and anyway reachable in a short period.

With reference to the storage extreme temperatures, it comes out:

- maximum T of incoming fluid in the charge phase. This value is restrained by the maximum reachable T outgoing the solar field, so that T_{max_in} = 175 °C is assumed;

- maximum T of outgoing fluid in charge phase. This value is restrained by the maximum sustainable T incoming the solar field. Hence the assumed value is the one for which, at equal ΔT in the collectors line, a mass capacity double to the nominal one is required, so $T_{max\_out} = 125\ °C$;

- minimum T of incoming fluid in discharge phase. This value is restrained by the water T outgoing from the exchanger supplying the power block in the working nominal conditions. Hence $T_{min\_in} = 80\ °C$;

- minimum T of outgoing fluid in discharge phase. This value is restrained by the minimum required fluid T incoming the exchanger supplying the ORC turbine. The imposed value is subsequently $T_{min\_out} = 130\ °C$.

It is to be noticed that the ΔT under which the extreme sections work is the same and it is equal to 45°C (Figure 6); additionally, considering the radiation conditions relative to a reference Italian site and the industrial nature of the typical user, $t_c = 1h$ has been assumed.

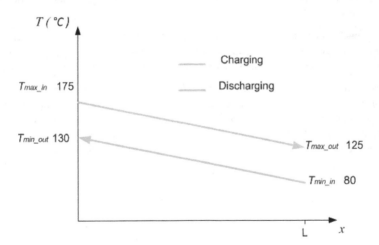

Figure 6. Restraints to HTF temperatures.

5.2. Definition of the physical model

A storage unit as the one considered here is composed by a piping bundle embedded within concrete for allowing fluid flow through it. The pipes, parallel one to the other, have axes (with reference to Figure 7) at a distance d_a. For the definition of the physical model, the same conditions have been assumed for each parallel channel, so that the storage module is considered to be composed by sub-units, named *elements*, put in parallel [10].

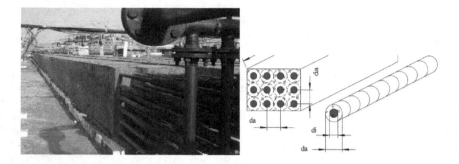

Figure 7. Typical storage module [3].

One element is composed by a concrete drilled cylinder with diameter d_a, in contact with a pipe with internal diameter d_i; the storage medium is characterized by thermal conductivity k, specific capacity per unit mass cp and density ρ. Even if the storage material inside contiguous cylinders is neglected, such a simplification is necessary for implementing an efficient algorithm.

By having this in mind, the problem of analysing the storage system moves from the study of the module to that of the single channel for which the initial section is the one of the incoming HTF during charge, whereas the final section is the outgoing one during the same phase.

By referring to an initial development stage for an innovative component, an analytical method is searched so that, via the application of a single formula or the implementation of a simple numerical procedure, it allows for reaching general evaluations on the main parameters involved in the design. Such a research stage is additionally adequate, with reference to the medium-term, considering the industrial potential of the Project.

In the following the main features of the procedure are described.

5.3. Analysis of the channel initial section

The initial section, located at the entrance of the hot fluid coming from the solar field, is the one subjected to the highest temperatures so resulting as the most critical one from the viewpoint of the thermo-mechanical design. In fact, by considering the humidity transport and dehydration phenomena effectively taking place in concrete (see e.g. [2, 11]) so that this section will be subjected to the highest thermal flux, it will sustain even the highest internal overpressures with the possibility of being exposed to spalling [12]; such a phenomenon is to be clearly avoided during first heating.

By following [13] with a distribution of piping bundles resembling an equilateral triangle (Figure 8), the attributable zone of each pipe far from the module external border is an hexagon which can be confused with the circumscribed circumference. On this circumference the fluxes can be practically considered as negligible and the entire analysis brought back to a single cylindrical channel with adiabatic external surface.

It is hence fully justified the necessity of a specific analysis for the considered section; in the following dimensionless parameters will be referred to, relative to the section only, as already occurs when studying *pebble bed* storage systems when dimensionless parameters such as porosity or specific exchange surface for unit length are fundamental.

To transfer the procedures adopted for pebble bed systems to embedded piping storage ones, a characteristic "porosity" of the section can be defined, determined by the area crossed by the fluid (not to be confused with the material one) in the following way (Figure 9)

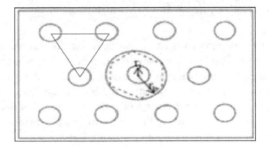

Figure 8. Location of the piping bundle and elementary cell [13].

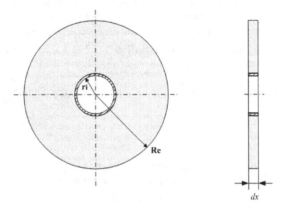

Figure 9. Geometrical parameters of the initial section.

$$\varepsilon = \frac{A_f}{A_{tot}} = \frac{\pi R_i^2}{\pi R_e^2} = \left(\frac{R_i}{R_e}\right)^2 \tag{4}$$

Such a porosity is obviously linked to the ratio between phases

$$\varepsilon = \frac{A_f}{A_{tot}} = \frac{A_f}{A_s + A_f} = \frac{1}{\dfrac{A_s}{A_f}+1} = \frac{1}{\dfrac{1}{f}+1} = \frac{f}{1+f}$$

(5)

And the specific exchange surface per unit length is defined

$$A_{sp} = \frac{2\pi\,R_i}{\pi\,R_e^2} = \frac{2R_i}{R_e^2} = \frac{2\varepsilon}{R_i}$$

(6)

A first consequence of adopting such an approach is that the storage module length becomes a parameter depending on the assumed section geometry, which is even physically justified by considering that the length is restrained by the production, at the end of the charge phase, of an outgoing flux at temperature not higher than the maximum sustainable one incoming the solar field. With reference to the charge phase it is clear that, for a given charging time, if the initial section does not reach a temperature distribution corresponding to the level assumed as complete charge, the same applies to the final sections so that the simulation of the entire channel is an useless effort.

Going into the specificity of the analysed section, the objective is to be able to estimate the relative *charging time* or, in a dual manner, the temperature level reached in correspondence of the external material circumference in a fixed interval of time. The charging time for a section is defined as the necessary time so that, at the corresponding abscissa, the temperature at the external concrete circumference of pertinence reaches a fixed value T_R^* at which the charge phase can be considered as concluded.

The interval of interest can be represented by the sum of two contributions

$$t_c = t_d + t_s$$

(7)

where t_d is the radial diffusion time necessary for heat to reach the external surface and t_s the rising time necessary for heat flux to increase the temperature T of that surface up to the fixed value.

As regards the former term, in literature the following estimate is proposed [14]

$$t_d \propto \frac{s^2}{\lambda} = \frac{s^2}{C_1 \cdot \lambda}$$

(8)

in which s is the material thickness subjected to heating and C_1 a proportionality coefficient linked to geometry.

The application to the given cylindrical geometry and the comparison with the results coming from the numerical procedure described in the following lead to

$$t_d = \frac{(R_e - r_i)^2}{13 \cdot \lambda} \tag{9}$$

Let's now examine the term related to the rising time; for doing this the following terms are introduced

$$C = \rho \, c_p \cdot \pi (R_e^2 - r_i^2) \, l \tag{10}$$

$$R = \frac{1}{2\pi \, r_i l \, h} + C_3 \frac{\ln\left(\frac{R_e}{r_i}\right)}{2\pi \, l \, k} \tag{11}$$

where C is the capacity term for the considered section, R the thermal resistance term given by the resistance due to the convective exchange on surface and by the solid internal conductivity, l the channel length assumed unitary, $C_3 = 0.72$ an empirical corrective coefficient calibrated from comparison with the results of the Finite Element model.

By observing that the product between the two above terms has the dimension of time,

$$t_1 = R \cdot C \tag{12}$$

and through the electrical analogy with resistive-capacitive circuits, it is immediate to recognize in t_1 a thermal charging time.

Whereas

$$T_R^* = C_2(T_{ing} - T_{ini}) + T_{ini} \tag{13}$$

in which C_2 is the fraction of engine ΔT assumed as sufficient to define the material as charged.

The analytical formula to calculate the charging time is

$$t_c = t_d - R \cdot C \cdot \ln\left(\frac{T_{ing} - T_R^*}{T_{ing} - T_{ini}}\right) \tag{14}$$

Equation (14), despite the presence of empirical calibration coefficients, is an analytical-type relation being derived from an exact physical model and it is pretty original. It is obtained by integrating the Fourier equation in the hypothesis of high Biot's numbers, giving an exponential-type trend for temperature

$$T(t)\big|_{r=R_e} = T_{ing} + (T_{ini} - T_{ing}) \cdot \exp\left[-\frac{t - t_d}{R \cdot C}\right] \tag{15}$$

where at the l.h.s. there is the temperature in correspondence of the external radius of the considered initial section, function of time.

By inverting Eq. (15) so to make time explicit, the above Eq. (14) is obtained.

The comparison between time histories of temperature given by Eq. (15), blue curve, and the numerical procedure, yellow curve, is depicted in Figure 10 with reference to a transient of 1 hour applied to a reference geometry.

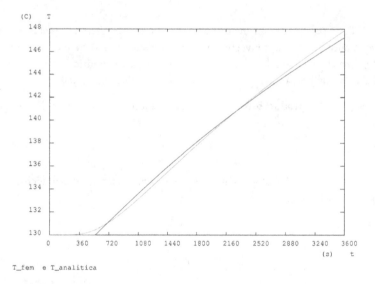

T_fem   e T_analitica

Figure 10. Validation of the analytical formula (Eq. (15)).

By varying the geometrical parameters, the difference between curves remains small, so that the obtained analytical formula is assumed to be validated. The formula, once T_R^* is fixed for a section, has the aim of predicting the time necessary to have this level reached, so the comparison must be performed at equal temperature evaluating the horizontal distance between curves, i.e. the difference between rising times from the two procedures to arrive at a given temperature. Additionally the resulting difference is not to be considered as an absolute value but relatively to the total charging time, that is a difference of 300 s with respect to an imposed charging time of 3600 s is fully acceptable in the initial design phase of the storage.

For sake of brevity the (similar) procedure for analysing the final channel section, as well as the Schumann analytical model (for pebble bed-type systems) [15] applied to this specific context have not been reported in this Chapter.

5.4. Application of the simplified design method

The development of a simplified design methodology for storage systems of the considered type represents a first goal for the developed research.

The starting point is an initial design for the plant; on the basis of these data the following operative scheme is followed.

i.   Channel section geometry. The geometrical parameters of the initial channel section (i.e. piping radius and external material radius) are defined (Figure 11).

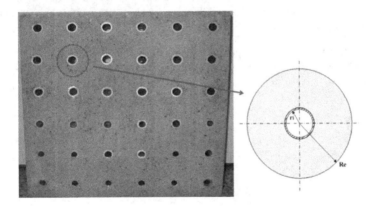

Figure 11. Initial section [3].

ii.   Calculation of the number of channels in parallel. Based on the hypothesized piping geometry and on fluid velocity and density, the capacity passing through the channel is defined. Being know the total plant capacity in nominal conditions, by dividing it to the determined unitary capacity, an estimate of the number of channels to be put in parallel is obtained to drain the whole flux.

iii.   Initial section analysis. Eq. (15) is applied for calculating the charging time of the initial section for evaluating the reachable entrance charge level.

$$T(t_c)\big|_{r=R_e} = T_{ing} + (T_{ini} - T_{ing}) \cdot \exp\left[-\frac{t_c - t_d}{R \cdot C}\right] = 170°C \qquad (16)$$

If during the imposed charging time it is not possible to charge the section at a fixed tempera-ture, even in the following sections such a level would not be reachable; on the basis of this sim-ple result it is hence necessary to change the assumed geometry and iterate phases II and III.

The fixed completed charge level is instead

$$T^* = 0,9\Delta T = 0,9\left(T_{ing} - T_{ini}\right) = 170,5°C \tag{17}$$

so that the hypothesized geometry allows for a practical complete charge of the initial section.

iv.   Channel length definition with graphical method. The section geometry is assumed as input for the Schumann's model.

a) A very long channel is hypothesized, L = 400 m.

Profiles of fluid temperature T_f are drawn along the channel for time instants up to the im-posed charging time and an horizontal line is superimposed, relative to the maximum value of T_f at the exit. The point in which the profile relative to t = t_c crosses this line is determined and the corresponding channel length value is read along the abscissa's axis (Figure 12).

The temporal trend of the fluid temperature is drawn on the final section with reference to the length of 300 m (Figure 13) and this value is checked against the maximum one, calculat-ing again the mean value at the final section.

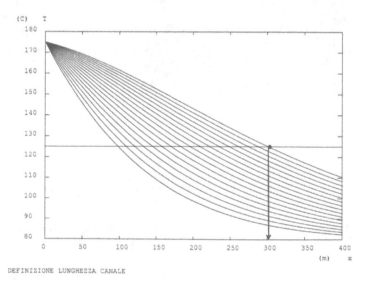

Figure 12. Channel length definition.

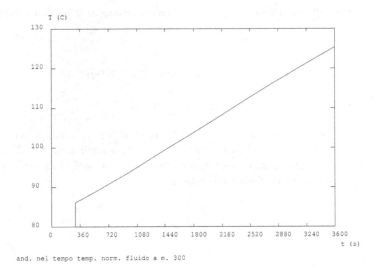

and. nel tempo temp. norm. fluido a m. 300

Figure 13. Temperature trend T_f.

v. Final section analysis. The calculated mean value allows for obtaining the temperature at the cold corner at the imposed charging time. Such a temperature is to be verified to correspond to the required charge level. It is to be noticed that in this section the condition of first heating is being evaluated.

vi. Finite Element simulation of the obtained configuration. Only at this stage, after having defined a possible module geometry via analytical formulas and with results in time of the order of about 1 min, a verification via FE models can be conducted so to even calculate the effective energy released to the material.

6. Numerical modelling

The following simplifications have been adopted:

- incoming fluid at constant temperature T_{ing} and capacity;

- material and fluid contained in the channel initially at constant temperature on the whole domain T_{ini}, that is conditions of first heating;

- the implemented algorithm for analysing the discharge phase assumes as initial temperature fields, both for fluid and solid, those calculated at the end of the charge phase;

- a fluid-dynamic analysis of the water flux within the channel is not conducted. The fluid system is treated via a 1D model whereas the solid domain is analysed via the FEM so that for each element the energy balance equation is solved (see e.g. [2]);

- concrete is modelled in a simplified manner as an homogeneous, continuous and isotropic material;

- the geometry is axis-symmetric.

Up to now all the procedures have been developed via CAST3M [16].

6.1. Development of a quasi-steady model

The first step is the implementation of a 1D model for the flux in the channel to be coupled with the FEM calculation. A quasi-steady procedure has been implemented in CAST3M on the basis of [13]. To validate the procedure, the results are compared to the reference ones. The physical channel model is shown in Figure 14.

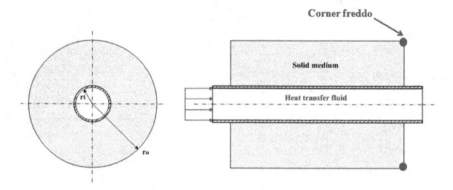

Figure 14. Channel geometry.

It consists in a concrete drilled cylinder with a fluid pumped through an internal pipe for heat exchange, with an adiabatic external surface ($r = r_0$). In order to solve the problem analytically, it is necessary to assume incoming velocity with a fully developed profile and negligible axial conduction in the solid medium.

The pipe has small thickness so that its thermal resistance is negligible; in this way the HTF is assumed to be in direct contact with the solid material.

The equation of the axial gradient for T_f is obtained by the energy balance on the channel element of length dz (Figure 15)

$$\rho c_p VA \cdot T_{f,in} - \rho c_p VA \cdot T_{f,out} - h \cdot 2\pi r_i dz \left(T_f - T_w\right) = \rho c_p Adz \cdot \frac{\partial T_f}{\partial t} \tag{18}$$

If neglecting the term of temporal variation for T,

$$\rho c_p VA \cdot T_{f,in} - \rho c_p VA \cdot T_{f,out} - h \cdot 2\pi r_i dz \left(T_f - T_w \right) = 0 \qquad (19)$$

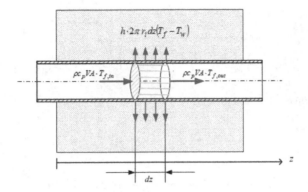

Figure 15. Energy balance for the fluid element.

where $A = \pi\, r_i^2$ is the crossed fluid section, V fluid velocity, $T_w = T_S(r_i)$ temperature of the channel wall.

By readjusting the terms

$$h \cdot 2\pi r_i \left(T_f - T_w \right) = \rho c_p VA \cdot \frac{\partial T_f}{\partial z} \qquad (20)$$

So that, by isolating the term of spatial variation of T_f and introducing the crossed fluid section,

$$\frac{2h}{\rho c_p V r_i} \left(T_f - T_w \right) = \frac{\partial T_f}{\partial z} \qquad (21)$$

Equation (21) is clearly non-linear for the presence of the unknown T_f on both sides; anyway, if for the term T_f within parenthesis a valid estimate can be given, the described relations reduce to linear differential equations of the first order which can be easily integrated so to give $T_f(z)$ at a specific time instant. In view of a step-by-step procedure in which a time discretization is performed, a valid estimate of T_f to be introduced in the l.h.s. is given by its value at the previous time-step.

To calculate the value of the known term, on the basis of the above discussed estimate of T_f, T_w values must be determined as well as the exchange coefficient h (via the Nusselt number) relative to the given conditions. As regards the former, it represents the output of the FEM calculation conducted on the domain corresponding to the zone occupied by concrete; for the latter, for turbulent fluxes in a cylindrical channel, the local value of the dimensionless parameter (oil and water) is given by

$$Nu = \frac{Re \cdot Pr \cdot \left(\frac{c_f}{2}\right)}{1.07 + 12.7 \cdot \left(Pr^{2/3} - 1\right)\sqrt{\frac{c_f}{2}}} \tag{22}$$

where the friction coefficient is given by

$$\frac{c_f}{2} = \left(2.236 \ln Re - 4.639\right)^{-2} \tag{23}$$

The range of validity of the above expression is $0.5 < Pr < 2000$, $10^4 < Re < 5 \times 10^6$.

By analysing the trend of the physical characteristic of liquid water and of the parameters used in the equation in which they appear (h and N_u), it can be noticed that the trends of N_u and h are weakly variable with temperature, as shown in Figure 16, so the choice of assuming constant values for these parameters and for water properties appears justified in the present model validation.

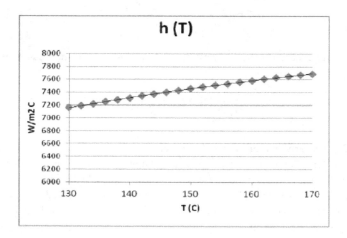

Figure 16. Variation of the thermal exchange coefficient of water with temperature.

After having implemented the numerical algorithm, the model characteristics relative to [13] and reported in Table 4 have been adopted.

r_i	r_o	L	T_{ing}	T_{ini}	ρ_s	c_{pS}	k_S	v_{H2O}
mm	mm	m	°C	°C	kg/m³	J/kg °C	W/m °C	m/s
13	65	2,6	90	25	2954	900	1	1

Table 4. Characteristics of the reference model.

Subsequently, a series of simulation runs have been launched by varying one of the listed parameters each time so to perform a sensitivity analysis; particularly thermal conductivity, external diameter and fluid velocity have been chosen. The analyses have allowed for obtaining time histories of temperature at the cold corner, i.e. at the concrete external circumference for the final section, representing the most hardly chargeable zone; the results have been then compared to the reference ones.

For sake of brevity, just the main observations are reported here:

- conductivity is evidently the parameter with a relevant influence on the storage charging mechanism considering that, by incrementing its value, temperature curves show higher slopes in correspondence of the considered zone; anyway, it is to be noticed that values around 5 W/m°C are physically unreachable with available concretes. Asymptotically, the results from the numerical code and those from the semi-analytical approach of [13] coincide;

- the system appears to be weakly sensitive to a variation in fluid velocity;

- the value of the internal radius has been maintained unaltered (that is the section crossed by the fluid and the characteristics of the convective exchange), so that the system capacity has been varied by changing the material thickness relative to the single pipe: the system appears to be strongly sensitive to a variation in the external radius.

6.2. Development of a non-steady model

The main weakness of the procedure described above is the difficulty linked to the number of iterations necessary to cover the entire channel length and to the integration of the various profiles; hence a new approach based on a non-steady model has been additionally developed.

A subdivision in axial cells of the channel has been realized so to follow the fluid along its path; the calculated temperature values are those related to the centre of the single cell.

The energy balance equation becomes (the capacity term is not neglected)

$$\rho c_p VA \cdot T_{f,i} - \rho c_p VA \cdot T_{f,u} - h \cdot 2\pi r_i dz \left(T_f - T_w\right) = \rho c_p A dz \cdot \frac{\partial T_f}{\partial t} \qquad (24)$$

By discretizing the time variable

$$\rho c_p V A \cdot \left(T_{f,j-1} - T_{f,j} \right) \Big|_{t-1} - h \cdot 2\pi r_i \Delta z \left(T_{f,j} - T_{w,j} \right) \Big|_{t-1} = \rho c_p A \Delta z \cdot \frac{T_{f,j}\big|_t - T_{f,j}\big|_{t-1}}{\Delta t} \qquad (25)$$

where j indicates the cell number, $T_{f,i} = T_{f,j-1}$ (T is evaluated at the centre of the cell and it is assumed, as incoming T in a cell, the value at the centre of the previous cell, Figure 17), t is the present time step and t-1 the previous one.

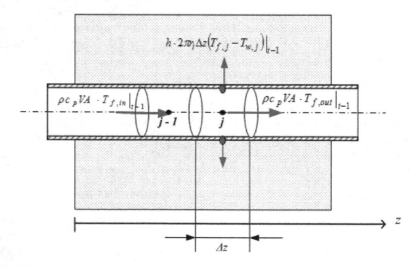

Figure 17. Definition of cells for the channel.

Hence T_f at the current t in the generic cell

$$T_{f,j}\big|_t = V \frac{\Delta t}{\Delta z} \cdot \left(T_{f,j-1} - T_{f,j} \right) \Big|_{t-1} - 2\pi \cdot r_i \Delta z \frac{h \cdot \Delta t}{\rho c_p A \Delta z} \left(T_{f,j} - T_{w,j} \right) \Big|_{t-1} + T_{f,j}\big|_{t-1} \qquad (26)$$

The time history of T_f and the final temperature field in the channel can be obtained step-by-step. Differently to the quasi-steady approach, this one allows for visualizing the progression of the hot front linked to the fluid flow within the pipe. The sequence of Figure 18 shows what stated with reference to a 100 m channel.

The hot fluid (red) enters the channel by increasing the wall temperature (blue) only up to the distance to which the first front has arrived, then the instant at which such front reaches the final section becomes evident and subsequently the profiles move up together until reaching the final configuration.

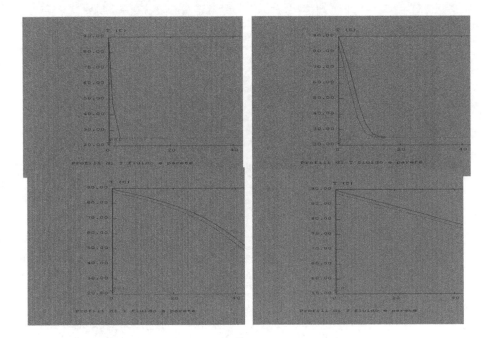

Figure 18. Fluid progression within the pipe.

Once again the approach has been validated against the results reported in [13] and the results have also been compared with the ones of the quasi-steady analysis; it can be immediately stated that for simulating a channel with limited length, the calculation times required by the present approach would be higher than those for the steady state (being possible to choose even large time-steps independently on the discretization of the grid). In fact a major restraint is now the time-step amplitude, being necessary that $\Delta t < \dfrac{\Delta z}{|v_{liq}|}$.

As an example, Figure 19 shows the time histories of temperature at the cold corner varying conductivity of the solid (1 and 3 W/m°C) and superimposing the present curve (black) with the ones from the quasi-steady analysis (blue) and from [13] (green): black and blue curves practically overlap, so that all validation issues (previously reported) apply even for the present approach. As an extension, for channels with length L < 10 m, the quasi-steady method allows for faster calculations -and susbstantially identical results- than the non-steady one. Hence it can be stated that the latter approach is efficient for analysing long channels, for which it is possible to assume Δz of the order of 1 ÷ 5 m and so even large Δts; under these assumptions even transients of about 1 h and channels of 100 ÷ 500 m can be solved in few minutes.

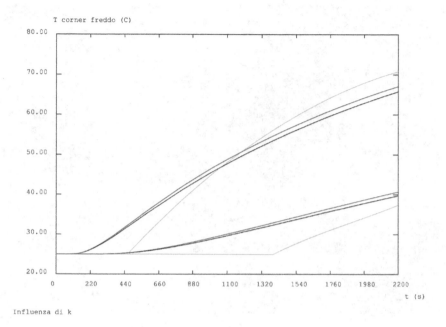

T corner freddo (C)

Influenza di k

Figure 19. Comparison among quasi-steady (blue), analytical (green) and non-steady models.

By assuming the same input data as before, a 1 h transient analysis has been developed for a 400 m channel (500 m is a typical value for modules from DLR) and taking
Δz = 2,5 m, Δt = 2 s; in Figure 20 the time histories of temperatures are depicted relatively to the external concrete circumference at the initial (yellow) and final (blue) section: the model follows the fluid in its own motion, so that temperature (initially fixed
at T_{ini} = 25 °C for the whole material) starts increasing on the external corner of the initial section after a time equal to the one necessary for the radial diffusion of heat; at
this time the hot front has not reached the cold one yet, which remains at the initial
temperature. The blue curve starts rising after a time equal to the sum between the
time of fluid transit in the channel and the one of radial diffusion of heat. Obviously,
the temperature at the initial section is, at equal time, higher than the one at the exit
section; additionally, considering the latter section, it comes out (the diagram has not
been added for sake of brevity) that the time histories of temperature for the fluid, for
the steel pipe surface and for the concrete in contact with the pipe substantially coincide (as predictable), so justifying the possibility of adopting models which neglect the
steel pipe and assume direct contact between fluid and concrete.

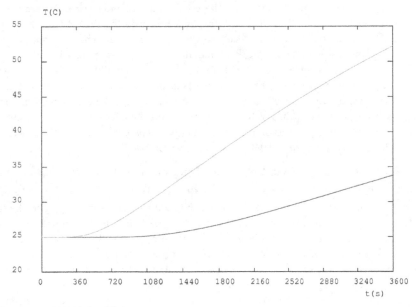

T corners esterni L= 400 m

Figure 20. Time history of temperature at extreme corners.

The non-steady model has been additionally used for simulating the discharge phase, not described here.

6.3. Energy issues

The FE code allows for even calculating the quantity of energy released by the fluid to concrete during the transient charge phase; with reference to the single channel of the examined configuration $E_{cls}=279MJ$. By multiplying this value by the number of channels, the total stored energy in the module becomes $E_{sto}=279MJ \cdot 44=12200MJ$, much lower to the 97200 MJ necessary to guarantee 6 hs storage. The situation is additionally worsened considering that the calculated energy value is overestimated with respect to the real one being referred to a charge cycle related to an initial field of uniform temperature within the solid equal to 80°C, so it is obtained by considering a ΔT higher than the effective one.

However, even if the obtained energy value is much lower than the one required during a preliminary plant design, it appears compatible with the formula for calculating the volume of material necessary to guarantee a fixed quantity of heat. In fact, for the considered geometry the volume results $V = N\pi\left(R_e^2 - r_i^2\right) \cdot L \approx 100m^3$ corresponding to about one tenth of the one which can be calculated on the basis of the nominal capacity of the plant system (see Table 3) [9]. The configuration has been consequently varied so to increase the storable heat quantity; the idea has been that to increase the number of channels in parallel, so it is necessary to decrease the single channel capacity to respect the plant restraint. The velocity of the flux has been so reduced from 1 m/s to 0.75 m/s leaving the geometry unaltered. By applying the simplified procedure, a channel length of 225 m is obtained for a total of 58 channels. Once checked that all the verifications conducted on the charge level and time are satisfied, a numerical simulation has been performed (final values for temperature on all sections have resulted to be the same as before). In this situation the energy released by the fluid to concrete for the single channel has been 209 MJ, so newly obtaining 12200 MJ.

Such a result was already predictable because generally true. In fact, if thermal losses towards the environment are neglected, in the charge phase $\Delta E_{cls} = \Delta E_{flu} = E_{f\_in} - E_{f\_out} \big|_0^{tc}$. By varying the capacity of the single channel, the total storage capacity, T_{in} and the trend of T_f on the final section (considering that the channel length is determined just imposing such a parameter) and the physical characteristics of the fluid remain unchanged, and so the last term of the expression above is unchanged. As a consequence, the storable solid energy is defined. Such a conclusion leads to formulate some observations on the material distribution within the storage module, that is on the convenience of having longer or shorter channels or, equivalently, if it is convenient to assume a high number of pipes in parallel.

The *comparison criterion* is that of the same material, i.e., as a first approximation, equal cost. Being the total capacity a parameter imposed by operational plant conditions, this must be the same whatever the number of modules in parallel; by having fixed the reference geometry for the transversal section, i.e. internal and external channel radius, and a charge time, two possibilities have been examined: 1) a channel with length L and fluid at velocity u; 2) two channels with length L/2 and fluid at velocity u/2 (the section is doubled).

The conclusion of this energy analysis of the charge phase has been that, at equal material and storable energy, a high number of channels in parallel with limited length is more convenient, i.e. with reduced velocities to reduce the charge losses along the pipes.

Such observations underline the potentialties of the developed procedure; in fact, even if it has not conducted to a satisfactory design of the storage module from the point of view of storable energy, it has anyway revealed:

• noticeable simplicity of application;

• reduced computational costs;

• reliability of produced results;

• capability of evidencing behavioural features of the component; particularly, it has been evidenced that the module storage capacity does not seem to represent a problem datum but a design consequence.

7. Conclusions

Guidelines for designing a concrete storage module and for its integration into a solar plant, respecting constraints linked both to an adequate solar field operation and to the production system based on ORC, have been described in this Chapter.

A series of simplified procedures have been developed to be used for a first module design and more sofisticated (even if more expensive) simulation techniques via the Finite Element Method have been checked and upgraded.

Once the ongoing experimental phase on a scaled storage prototype at the ENEA site of Casaccia has been concluded, the obtained data will be used for completing both the setup of the calculation instruments and the R&D activity dealing with the development of an appropriate concrete mixing, optimizing its chemical-physical and durability performances, and with the module integration within a CSP system.

Acknowledgments

The research work is partly funded by the Fondazione Cassa di Risparmio di Trento e Rovereto, Prot. SG 2483/10.

Author details

Valentina A. Salomoni[1*], Carmelo E. Majorana[1], Giuseppe M. Giannuzzi[2], Rosa Di Maggio[3], Fabrizio Girardi[3], Domenico Mele[1] and Marco Lucentini[4]

*Address all correspondence to: valentina.salomoni@dicea.unipd.it

1 Department of Civil, Environmental and Architectural Engineering, University of Padua, Padua, Italy

2 ENEA – Agency for New Technologies, Energy and Environment, Thermodynamic Solar Project, CRE Casaccia, Rome, Italy

3 Department of Materials Engineering and Industrial Technologies, University of Trento, Trento, Italy

4 CIRPS, University of Rome "La Sapienza", Rome, Italy

References

[1] Giannuzzi G.M., Majorana C.E., Miliozzi A., Salomoni V.A., Nicolini D. Structural design criteria for steel components of parabolic-trough solar concentrators. Journal of Solar Energy Engineering 2007;129 382-390.

[2] Salomoni V.A., Majorana C.E., Giannuzzi G.M., Miliozzi A. Thermal-fluid flow within innovative heat storage concrete systems for solar power plants. International Journal of Numerical Methods for Heat and Fluid Flow 2008;18(7/8) 969-999.

[3] Laing D., Steinmann W.D., Tamme R. Solid Media Thermal Storage for Parabolic Trough Power Plants. Solar Energy 2006;80 1283–1289.

[4] Laing D., Lehmann D., Fiss M. Test Results of Concrete Thermal Energy Storage for Parabolic Trough Power Plants. Journal of Solar Energy Engineering 2009;131(4).

[5] Salomoni V.A., Majorana C.E., Giannuzzi G.M., Di Maggio R., Girardi F., Brunello P. Conceptual study of a thermal storage module for solar power plants with parabolic trough concentrators. In: Ubertini F, Viola E, de Miranda S, Castellazzi G (eds.) XX National Congress AIMETA, 12-15 Sept. 2011, Bologna, Italy. Conselice (Ra): Publi&Stampa; 2011.

[6] Di Maggio R., Ischia G., Bortolotti M., Rossi F., Molinari A. The microstructure and mechanical properties of Fe-Cu materials fabricated by pressure-less-shaping of nanocrystalline powders. Journal of Materials Science 2007;42 9284-9292.

[7] Laing D., Balh C., Bauer T., Lehmann D., Steinmann W.D. Thermal Energy Storage for Direct Steam Generation. Solar Energy 2011;85(4) 627–633.

[8] Tamme R., Laing D., Steinmann W.D. Advanced Thermal Energy Storage Technology for Parabolic Trough. ASME Journal of Solar Energy Engineering 2004;126 794-800.

[9] Santoro M. Study and analysis of thermal storage systems in cementitious materials for small size concentrated solar power plants. Master Thesis. ENEA-University of Rome "La Sapienza"; 2012.

[10] Sragovich D. Transient analysis for designing and predicting operational performance of a high temperature sensible thermal energy storage system. Solar Energy 1989;43(1) 7–16.

[11] Majorana C.E., Salomoni V, Schrefler B.A. Hygrothermal and mechanical model of concrete at high temperature. Materials and Structures 1998;31 378-386.

[12] Majorana C.E., Salomoni V.A., Mazzucco G., Khoury G.A. An approach for modeling concrete spalling in finite strains. Mathematics and Computers in Simulation 2010;80(8) 1694-1712.

[13] Bai F., Wang Z., Liye X. Numerical Simulation of Flow and Heat Transfer Process of Sol-id Media Thermal Energy Storage Unit. In Goswami D, Zhao Y (eds.) Solar Ener-

gy and Human Settlement: proceedings of ISES Solar World Congress 2007 (Vol. I-V). Springer; 2009;10 2711-15.

[14] Faghri A., Zhang Y. Transport Phenomena in Multiphase Systems. Burlington: Elsevier Inc.; 2006.

[15] Schumann T.E.W. Heat transfer: a liquid flowing trough a porous prism. Journal of the Franklin Institute 1929;208(3) 405–416.

[16] CAST3M Users' Manual, http://www-cast3m.cea.fr/.

Chapter 6

Physical and Technological Aspects of Solar Cells Based on Metal Oxide-Silicon Contacts with Induced Surface Inversion Layer

Oleksandr Malik and F. Javier De la Hidalga-W

Additional information is available at the end of the chapter

1. Introduction

With the current concerns about the worldwide environmental security, global warming, and climate change due to the emission of CO_2 from the burning of fossil fuels, it is desirable to have a wide range of alternative energy technologies. Photovoltaic, or solar cells, have already proven themselves to be a viable option as a nonpolluting renewable energy source, as well as a visible business that will grow stronger in the global economy of present and future centuries.

The main problems of global practical application of solar cells for energy production are their low efficiency (typically of about 10-15%) and the cost of photovoltaic modules ($200-500/m^2). Monocrystalline silicon is the main material for the fabrication of solar cells. It is the most studied material, and the lifetime of silicon solar cells and modules is 15-30 years. The higher efficiency of silicon solar cells (up to 24%) that can be achieved using a complicated cell design, and applying new technological processes, lead to an undesirable increase of their total cost. From this point of view, solar cells based on more simple Schottky contacts and metal-insulator-semiconductor (MIS) structures are promising for solar energy conversion due to their relatively low production cost.

Since 1978, a new class of photovoltaic devices, namely the semiconductor-insulator-semiconductor (SIS), has emerged, using a deposited conductive top layer made of a degenerated wide-bandgap oxide semiconductor. Excellent results have been reported using tin-doped indium oxide (In_2O_3:Sn or ITO). Other oxide semiconductors, such as fluorine-doped tin oxide (SnO_2:F) and doped zinc-oxide (ZnO), have also been used as a transparent conducting electrode.

Thin films of these oxides behave as a metal, thus such SIS structures present electrical properties similar to those presented by MIS devices. Of course, the optical and photoelectrical properties of SIS structures exceed the properties of MIS devices.

In our chapter of the book "Solar Energy", edited by Radu D. Rugescu [1], the reader can find a complete bibliography regarding the SIS solar cells and the properties of transparent conducting oxides fabricated using different methods.

In that book, a preceding discussion regarding the fabrication process of SIS solar cells, the structural, electrical and optical properties of ITO and SnO_2:F thin films, as well as the physical model of spray deposited ITO-Si solar cells and theirs properties has been presented. It was shown that such structures present a high barrier height that is not typical for Schottky diodes. The authors developed a physical model of the ITO-Si solar cells based on an inversion p-n junction similar to that reported by J. Shewchun et al. [2] for MIS structures with an Al electrode. According to this model, the I-V characteristics are dominated by a diffusion current flow in the bulk of the silicon substrate and show the usual behavior for a Shockley diode.

The aim of this chapter is to discuss some new physical aspects of spray deposited ITO-Si solar cells which are tightly connected with the fabrication technology. Below, we will show that suitable process schedules for chemical treatment of the silicon surface with the presence of acceptor-type surface states, is the reason for the inversion of the conductivity type at the silicon surface. A sufficiently high potential barrier can be formed *before* the deposition of the ITO film if a minimum amount of fixed charge appears within the interfacial layer very close to the silicon surface. Then the role of the ITO electrode is the formation of an ohmic contact on the inversion layer.

2. Barrier height of MIS solar cells

Solar cells based on contact metal-semiconductor with a Schottky barrier really represent MIS structures due to the existence of a thin insulator layer between the metal and the semiconductor. MIS solar cells are receiving increasing attention because they present several inherent advantages such as low cost, high yield, fabrication at low substrate temperature, etc. However, one drawback of such cells is that their open-circuit voltage is slightly low and depends on the potential barrier height.

The expression to calculate the barrier height ϕ_{Bn} for n-type substrates is given by [3]

$$\phi_{Bn} = \gamma\left(\phi_m - \chi_s\right) + \left(1-\gamma\right)\left(E_g - \phi_0\right) - \left(\frac{1-\gamma}{D_s}\right)\frac{Q_{ox}}{q} \tag{1}$$

where

$$\gamma = \left(1 + \frac{q\delta D_s}{\varepsilon_i}\right)^{-1} \qquad (2)$$

Here, ϕ_m is the metal work function, χ_s is the semiconductor electron affinity, E_g is the semiconductor energy gap, ϕ_0 is energy level of surface states at the semiconductor surface, δ is insulator (oxide) thickness, D_s is density of surface states, and $\frac{Q_{ox}}{q}$ represents the amount of fixed charge lying within the interfacial layer very close to the insulator-semiconductor interface; the other symbols have their usual meaning. Figure 1 shows the calculated variation of the barrier height with the metal work function for different values of interfacial layer thickness [3]

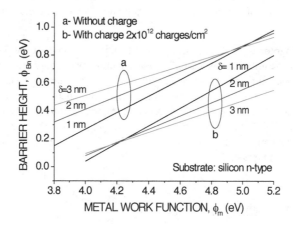

Figure 1. Calculated variation of the barrier height with metal work function for different values of the interfacial layer thickness [3].

These calculations were conducted assuming D_s=3x10^{12} states/cm²/eV and ϕ_0=0.27 eV for silicon. Positive oxide (insulator) charge was fixed to 2x10^{12} charges/cm². The positive sign of the charge is typical for several chemical methods used for the fabrication of thermal thin oxide on the silicon surface. It is clear that the existence of this charge decreases the barrier height even if the metal work function is as high as 5.2 eV; for this case, the barrier height does not exceed 0.8 eV.

Another situation becomes apparent when the fixed charge is negative, for example, in presence of acceptor-like surface states. In this case the sign of the third term in equation (1) is positive, and a higher value of the barrier height is possible. Thermally grown oxide on the silicon surface usually presents a positive fixed charge, however it may be possible to intro duce a negative charge in the oxide by developing suitable process schedules for the chemical treatment of the semiconductor surface.

3. Relation between the charge and surface potential

For our discussion, we must find a relationship between the charge on the semiconductor surface Q_s and the surface potential ϕ_s. Considering a semiconductor having donor and acceptor impurities N_d and N_a, respectively, completely ionized at room temperature, we need to solve the Poisson equation

$$\frac{d^2\varphi}{dx^2} = -\frac{4\pi\rho}{\varepsilon}; \quad \rho = q[N_d - N_a + p(x) - n(x)].$$

Using the boundary conditions:

$$\varphi = \varphi_s \lhd 0 \quad at\, x = 0,$$

$$\varphi \to 0, \quad d\varphi/dx \to 0 \quad at\, x \to \infty.$$

In the semiconductor volume we consider charge neutrality:

$$N_d + p = n + N_a,$$

where p and n are the concentrations of electrons and holes, respectively, and

$$n(x) = n\exp(q\varphi/kT), \quad p(x) = p\exp(-q\varphi/kT).$$

Defining $n/n_i = n_i/p = \gamma$ and $\psi = q\varphi/kT$.

Then the Poisson equation can be written as

$$\frac{d^2\psi}{dx^2} = -\frac{4\pi q^2 n_i}{\varepsilon kT}[\gamma(1 - \exp\psi) + \gamma^{-1}(\exp(-\psi) - 1]$$

After integrating both sides with respect to ψ and determining the integration constant from the boundary conditions: $\psi \to 0$ and $d\psi/dx \to 0$ $at\, x \to \infty$

$$\left(\frac{d\psi}{dx}\right)^2 = L_D^{-2}[\gamma(\exp\psi - 1) + \gamma^{-1}(\exp(-\psi) - 1) + \psi(\gamma^{-1} - \gamma)]$$

Where $L_D^{-2} = 8\pi q^2 n_i/\varepsilon kT$, and n_i is the intrinsic carrier concentration.

It is easy to obtain a differential equation for the potential ϕ in the form

$$\frac{d\phi}{dx} = -(kT/qL_D)\left[\gamma\left(\exp\left(\frac{q\phi}{kT}\right) - 1\right) + \gamma^{-1}\left(\exp - \left(\frac{q\phi}{kT}\right) - 1\right) + \frac{q\phi}{kT}(\gamma^{-1} - \gamma)\right]^{1/2}$$

At $x=0$, the boundary condition is $\varepsilon E_{x=0} = 4\pi Q_s$,

Where $E_{x=0} = -d\phi/dx$, and $E_{x=0}$ is electric field at the surface of the semiconductor.

Finally, we obtain

$$Q_s = 2qn_iL_D \left[\gamma(\exp\left(\frac{q\varphi_s}{kT}\right) - 1) + \gamma^{-1}(\exp-\left(\frac{q\varphi_s}{kT}\right) - 1) + \frac{q\varphi_s}{kT}(\gamma^{-1} - \gamma) \right]^{1/2} \qquad (3)$$

Here, $\varphi_s \vartriangleleft 0$, and Q_s is a positive charge. For an n-type semiconductor, a band bending will be developed due to the accumulation of electrons at the semiconductor surface.

For a negative charge Q_s, we need to change the sign of φ_s in equation (3).

Figure 2 shows the calculated dependences of the negative Q_s on the surface potential at the silicon surface for different concentrations of donors in the silicon substrate.

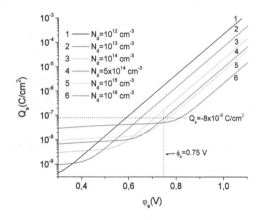

Figure 2. Calculated dependence of Q_s on the surface potential at the silicon surface for different donor concentrations in the silicon substrate.

We will discuss the properties of solar cells fabricated on n-type silicon with $N_d=5\times10^{14}$ cm^{-3}. If a negative charge $Q_s=8\times10^{-8}$ C/cm^2 is present on the Si surface, the surface potential 0.75 eV is due to the strong inversion condition because $q\varphi_s > 2(E_F - E_i)=0.28eV$, where E_F and E_i are the Fermi and intrinsic energy levels in the silicon substrate, respectively.

4. Surface potential of the silicon wafers after a chemical cleaning

For any technical application, the silicon wafer must be subjected to a certain schedule of chemical treatment. These are very important technological steps necessary to prevent the contamination of the future device from non-controlled sources such as some metals that can introduce deep energy levels into the substrate.

The RCA Standard Cleaning, developed by W. Kern and D. Puotinen in 1965, and disclosed in 1970, is extremely effective for removing contamination from silicon surfaces, and it is a current industry standard. The RCA cleaning has two sequential steps: the Standard Cleaning 1 (SC-1), followed by Standard Cleaning 2 (SC-2): The SC-1 solution, consisting of a mixture of ammonium-hydroxide (NH_4OH), hydrogen-peroxide (H_2O_2), and water (H_2O), is the most efficient particle removing agent found to date. This mixture is also referred to as the Ammonium-Hydroxide/Hydrogen-Peroxide Mixture (APM).

In the SC-1 cleaning solution, the formation of native oxide (SiO_2) by hydrogen peroxide and the etching of the silicon oxide by alkalis (NH_4OH) operate simultaneously. For this reason, the overall wafer thickness is slowly reduced, but certain thickness of SiO_2 forms on the wafer surface. In the past, the SC-1 solution presented a tendency to deposit metals on the surface of the wafers, and consequently a treatment with the SC-2 mixture (H_2O: HCl: H_2O) were necessary to remove metals.

It is possible to consider SC-1 and SC-2 solutions as agents for *wet chemical oxidation* of the silicon wafer due to the formation of a thin silicon dioxide layer during the treatment of the wafer in these solutions. Other chemical agents such as HCl, HNO_3, the H_2SO_4:H_2O_2 mixture, hydrogen-peroxide (H_2O_2), and water (H_2O), can also be used for wet chemical oxidation. With a suitable thickness and physical parameters, these oxide layers can be used for the fabrication of MIS solar cells. However, the question is what the thickness of the silicon oxide layer will be obtained and which surface potential on the wafer will be developed after applying this technological procedure? One more question is connected with the density of surface states after the wet chemical oxidation.

Many researchers have tried to find the answer to these questions. For instance, a complete investigation on this issue was conducted by Angermann [4]; some parameters of the oxide layers formed with different chemical agents are shown in Table 1.

Oxidizing solutions	Composition	T [^{0}C]	Treatment time (min)	$D_{it,min}$ [x10^{12} cm^{-2}eV^{-1}]	$<d_{ox}>$ [nm]
SC-1	6:1:1	75	10	6	1.1
SC-2	5:1:1	75	10	5	1.3
SC-1+SC-2	-	75	10+10	6	1.2
H_2SO_4:H_2O_2	1:1	120	5	5	1.8
HCl	36%	40	5	3	-
HNO_3	65%	60	5	8	-
Deonized H_2O	18 MΩ-cm	80	120	0.4	1.5 Si (111) 2.5 Si (100)

Table 1. Parameters of the oxide layer formed with different chemical agents [4], $D_{it,min}$ is the minimum density of surface states.

Other published results [5, 6] present a thickness of the oxide in the range of 0.8-1 nm after treatment in SC-1 solution. From Table1 it is clear that the best result regarding the minimum density of surface states is obtained by using hot water. With other chemical agents $D_{it/min}$ exceeds the value of 10^{12} cm^{-2}eV^{-1}. After etching in NH$_4$F during 6.5 min, the oxide thickness obtained with the SC-1 solution decreases to 0.3 nm, and the density of surface states is 1x10^{11} cm^{-2}eV^{-1}. Such parameters are suitable for the fabrication of Schottky diodes based on metal-semiconductor structures.

Now, it is interesting to know the band bending of the silicon surface after different processes for obtaining the chemical grown oxide. Again, and according to reference [4], the position of the Fermi level $E_{Fs}=E_F-E_i$ (at x=0), determined at the n-type Si surface, with bulk Fermi level $E_{fb}=E_F-E_i$ (x>0) =-0.32 eV, after HF and NH$_4$F and subsequent wet chemical oxidation in various solutions, is shown in table 2.

Chemical etching Wet chemical oxidation agent							
HF	NH$_4$F	SC-1	SC-2	H$_2$SO$_4$/H$_2$O$_2$	HCl	HNO$_3$	H$_2$O
+0.32	-0.02/+0.1	-0.25	+0.1	-0.03	+0.18	+0.2	≅0

Table 2. Fermi level position $E_{Fs}=E_F-E_i$ (at x=0) determined on n-type Si surfaces when the bulk Fermi level is $E_{fb}=E_F-E_i$ (x>0) =-0.32 eV, after HF and NH$_4$F, and subsequent wet chemical oxidation in various solutions.

The HF treatment leads to a strong inversion layer on H-terminated p-type silicon surfaces, which results from a positive charge induced by electronegative surface groups (-H, -O-H, and –F) on the surface silicon atoms. Using NH$_4$F as the final etching agent under cleanroom conditions, the remaining amount of surface charges results from the electro-negativity difference between silicon and hydrogen.

After the wet-chemical oxidation of initially H-terminated surfaces, characteristic values of the surface Fermi-level E_{Fs}, as shown in Table 2, were obtained from the interface-trapped charge and from different kinds of oxide charges.

Most of the oxidizing solutions SC-2, HCl, and HNO$_3$ cause a strong depletion of holes on p-type silicon surfaces due to the positive fixed oxide charge, which is also known from CV measurements of thermally grown oxides.

In contrast, the SC-1 process causes a negative surface charge, which is possible to originate from the dissociation of ≡Si-OH groups of the oxide layer in the alkaline solution (≡Si-OH ⇔ ≡Si-O$^-$+H$^+$).

From this reported results we make an important conclusion regarding the use of an n-type silicon substrate: the forming of transparent for carriers insulating layer after wet oxidation and the formation of a depletion or inversion layer on the silicon surface after substrate treatment in the SC-1 solution. All other treatments in wet oxidizing solutions will produce an accumulation band bending.

5. Chemical oxide after treatment in hydrogen-peroxide

The chemical oxide can also be created on the silicon surface with the treatment of the wafer in an aqueous solution of hydrogen-peroxide. Neuwald et al. [7] shown that a very thin (about 0.5 nm) oxide is formed after immersion of a (111) silicon wafer in ultrapure 30% H_2O_2 solution. XPS analysis shows that the oxide thickness saturates on this level after 10 minutes of immersion in the solution. Other results published by Verhaverbeke et al. [8] discuss in detail the limitation of the oxide thickness obtained in H_2O_2 and SC-1 solutions. They show that the oxide thickness obtained in H_2O_2 solutions for different concentrations as function of time does not exceed 0.9 nm. In order to explain this experimental fact they used the results of Stoneham and Tasker, where the effect of image charges and their influence on the grown oxide films are studied. These authors found that the polarization energies associated with localized charges near the interface between oxides and silicon provide a driving force, over short distances, which affect the transport of peroxide anions HO_2^- (principal oxidant) to the silicon surface. As the oxide thickness grows, the image charge reduces the transport of the ionic species, and the oxidation process is limited. According to Verhaverbeke, the model based on the charge transport (Figure 3) that predicts the frequently observed limitation of an oxide thickness of around 0.8-1 nm, may also be applied to the oxide grown in SC-1 solutions with a certain content of hydrogen-peroxide. It is not possible to know in advance the sign of the charge in the oxide formed with ultrapure H_2O_2, but below we show that the presence of some impurities in the hydrogen-peroxide solution can change drastically the situation.

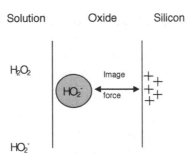

Figure 3. Schematic illustration of Image Charge transport in H_2O_2 and SC-1 solutions [8].

6. Chemical oxide fabricated with a contaminated SC-1 solution

Generally, silicon wafers always present some charge in either, chemically or thermally grown native oxide. It is well known that a fixed positive charge appears on thermally oxidized Si wafers. It is also known that a large positive fixed oxide charge appears in naturall oxidized Si wafers soon after they are dipped in an aqueous hydrofluoric acid (HF), and then the charge decreases as the native oxide grows in air. Munakata and Shimizu [9,10] reported that when silicon wafers are rinsed by an SC-1 solution, a fairy large negative charge of 5.8×0^{11} charges/cm^2 is observed in commercially available n-type Si wafers. In this case, the wafers were rinsed with the SC-1 solution in a Pyrex glass container. The charge was significantly smaller when the treatment was conducted using a quartz container; this is because the Al (and also Fe) concentration in quartz is generally more than one order of magnitude lower than that found in Pyrex glass. This fact clearly suggests that some species in the SC-1 solution must be the cause of the negative charge, whose density should be much higher than that of the positive fixed oxide charge. At this moment, the exact chemical mechanism of the negative charge formation is not clear. Authors assumed that the negative charge arises from (AlSiO)⁻ networks, when 3-valence Al substitutes 4-valence Si in the oxide. The role of Al to form the negative oxide charge on the Si surface after rinsing with the SC-1 solution was verified with especially Al-contaminated SC-1 solutions. The same results were also obtained using 3-valence iron (Fe)–contaminated solutions.

It is known that Al atoms can isomerically substitute the Si atoms forming a wide class of $(SiO_2)_x \times (Al_2O_3)_y$, and Al can penetrate in the SiO$_2$ film at a depth of about 100 nm. The Al atoms in the SiO$_2$ film break partially or completely the dπ -$\pi\sigma$ bounds. In three-coordinated states, these atoms are strong acceptors of electrons.

We assume that the negative charge can also appear in oxides obtained using Al (or Fe)-contaminated hydrogen-peroxide (H$_2$O$_2$) solutions.

7. Work function of tin-doped indium oxide (ITO) films

In this section we discuss the work function ($q\phi_{ITO}$) of the ITO film. The data about the value of $q\phi_{ITO}$ is not presented systematically in the literature. This parameter presents a strong dependence on the fabrication method, structure and morphology of the film, and also on the carrier concentration. Reported results for films obtained by thermal evaporation give a work function of 5.0 eV [11]. The work function of pure In$_2$O$_3$ films obtained by RF magnetron sputtering was found to be in the range of 5.3-5.4 eV [12], and in contrast, the work function of ITO films with 5-20 wt. % Sn was found in the 4.6-4.8 eV range.

The work function of the ITO films fabricated by pyrosol technique was reported around 4.8 eV, without an additional thermal treatment, whereas it was around 5.2 eV after a thermal treatment [13].

In the case of n-type degenerated semiconductors, such as the ITO films, the work function is expected to shift when the carrier concentration n_e changes. It is inversely proportional to $n_e^{2/3}$; nevertheless, an opposite trend has been found for the relationship between the optical energy gap E_g^{opt} and n_e. The Fermi energy in the film conduction band should vary for an increasing or decreasing n_e. Therefore, the control of the work function with n_e in the ITO layers is an important issue to take into account for the estimation of the energy barriers in

Schottky and MIS devices. According to Sato et al. [14], the optical band gaps of the ITO films deposited by dc magnetron sputtering increased from 3.8 to 4.3 eV when the carrier density increased from 8.8×10^{19} to 8.2×10^{20} cm^{-3}, whereas the work function decreased from 5.5 to 4.8 eV. The variation of the optical band gap (E_g^{opt}) and the work function (ϕ), as a function of the two-thirds power of the carrier density ($n_e^{2/3}$) for undoped In$_2$O$_3$ and ITO films, is shown in Figure 4.

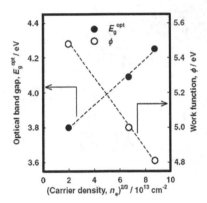

Figure 4. Variation of the optical band gap (E_g^{opt}) (solid circles) and the work function (ϕ) (open circles) as a function of the two-thirds power of the carrier density ($n_e^{2/3}$) for undoped In$_2$O$_3$ and ITO films [14].

For our solar cell we are using Sn-doped indium oxide films (ITO) fabricated by spray pyrolysis technique. The optimized films have a carrier concentration of about 10^{21} cm^{-3}, and the shift of the Fermi level is about 0.5 eV. Taking into account the reported data about the work function, we will use $q\phi_{ITO} = 4.8$ eV.

8. Fabrication of ITO and FTO films by spray pyrolysis

The spray pyrolysis technique was employed for the deposition of thin ITO and FTO films on glass and sapphire substrates in order to investigate their structural, electric, and optical properties. A 10 Ω-cm n-type (100)-silicon substrate, chemically cleaned and specially treated, was used for the fabrication of solar cells. The apparatus (atomizer) for the spray deposition was designed for obtaining small-size droplets. The substrates were mounted on a heater covered with a carbon disc for obtaining uniform temperature. Spraying was made using compressed air. Periodical cycles of the deposition with duration of 1 sec and intervals of 5 sec were employed to prevent a rapid substrate cooling. The deposition rate was high, of about 200 nm/min. For the ITO films deposition, 13.5 mg of InCl$_3$ were dissolved in a 170 ml mixture of ethylic alcohol and water in a 1:1 proportion, and adding 5ml of HCl. The different ratios of Sn/In achieved in the ITO films were controlled by adding in the solution a calculated amount of tin chloride (SnCl$_4$*5H$_2$O). The substrate temperature, in the range of

380-480°C, was controlled using a thermocouple with an accuracy of ±5°C. The optimum distance from the atomizer to the substrate and the compressed air pressure were 25 cm and 1.4 kg/cm², respectively. Figure 5 shows schematically the equipment set-up.

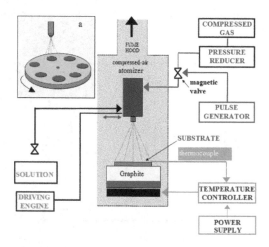

Figure 5. Set-up of the equipment used for the deposition of the ITO films. Insert (a) shows the modification of the equipment for the purpose of mass-production.

9. Characterization equipment and methods

The film thickness was measured with an Alpha Step 200 electronic profilometer. The electrical resistivity, Hall mobility and carrier concentration were measured at room temperature using the van der Pauw method. Hall effect parameters were recorded for a magnetic field of 0.25 Tesla. The optical transmission spectrum was obtained using a spectrophotometer. The structural characterization was carried out with an X-ray diffractometer operating in the Bragg-Brentano Θ-Θ geometry with Cu K_α radiation. A JSPM 5200 atomic force microscope was used to study the film surfaces. The chemical composition of the films was determined using an UHV system of VG Microtech ESCA2000 Multilab, with an Al- K_α X-ray source (1486.6 eV) and a CLAM4 MCD analyzer.

10. Properties of the spray deposited ITO films

The X-ray diffraction (XRD) measurements shown in Figure 6 indicate that all deposited ITO films, with thicknesses in the 160-200 nm range, and fabricated from the chemical solutions for different Sn/In ratios, presents a cubic bixebyte structure in a polycrystalline configuration with (400) as the preferential grain orientation

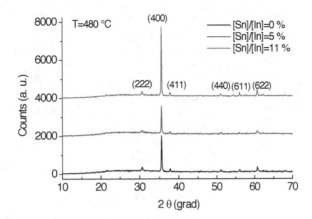

Figure 6. XRD spectra of the ITO films fabricated using precursors with different Sn/In ratios. The mean size of the grains, 30-50 nm, was determined using the classical Debye-Scherrer formula from the half-wave of the (400) reflections of the XRD patterns.

A surface roughness of about 30 nm was determined from images of the film surfaces obtained with an atomic force microscope (Figure 7).

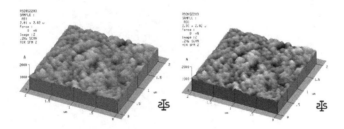

Figure 7. Atomic force images of the In_2O_3 film (left) and the ITO film with 5% Sn/In (right).

Figures 8 and 9 show the dependence of the electric parameters of deposited ITO film on the ratio Sn/In used in the solution for the fabrication of the films. The sheet resistance R_s shown in Figure 8 has a minimum 12 $\Omega/\square$ for the films prepared using a solution with 5% Sn/In ratio.

The minimum resistivity obtained for the films deposited with a 5% Sn/In solution is 2×10^{-4} Ω-cm. The variation of carrier concentration as a function of the Sn/In ratio in the precursors is shown in Figure 9. The carrier concentration presented a value of 1.1×10^{21} cm^{-3} at the 5% Sn/In ratio. This high value is comparable with better results achieved when other technological methods are used for the fabrication of the ITO thin films.

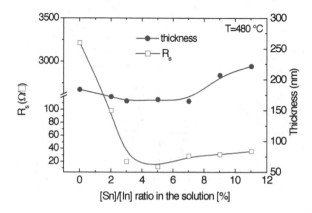

Figure 8. The sheet resistance as a function of the Sn/In ratio in the precursor used for the film deposition. The thicknesses of the films are also shown.

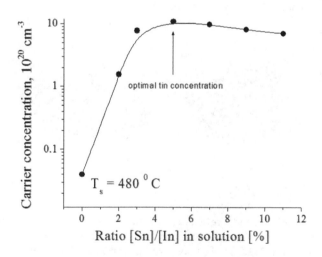

Figure 9. Variation of the carrier concentration as a function of the Sn/In ratio.

The optical transmission of indium oxide films for two thicknesses deposited on glass substrates, as a function of the wavelength, is shown in Figure 10.

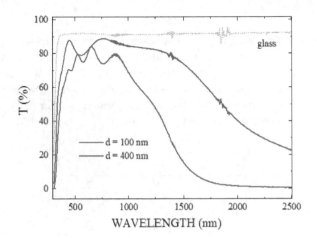

Figure 10. Optical transmission of the ITO films for two thicknesses as a function of the wavelength

The use of sapphire substrates allows for determining the optical energy gap of the ITO films by the extrapolation of the linear part of the $\alpha^2(h\nu)$ curves to $\alpha^2=0$, where α is the absorption coefficient. The optical gap increases with the carrier concentration due to the well known Moss-Burstein shift. For ITO films fabricated using the solution with a 5% Sn/In ratio, this shift is 0.5 eV, and the optical gap is 4.2 ± 0.1 eV. Such high value of the optical gap offers transparency in the ultraviolet range, which is of fundamental importance in solar cell applications. Because of the opposite dependence of conductivity (σ) and transmission (T) of the ITO film on its thickness (t), both parameters must be optimized. A performance comparison of different films is possible using $\phi_{TC}=T^{10}/R_s=\sigma t\exp(-10\alpha t)$ as a figure of merit [21]. Table 3 compares the values of ϕ_{TC} for spray deposited ITO films reported in this work with some results obtained by other authors but using a different deposition technique.

Process	R_s, $\Omega/\square$	T (%)	ϕ_{TC} (Ω^{-1}) $\times 10^{-3}$	Reference
spray	26.0	90	13.4	[15]
spray	9.34	85	21	[16]
spray	10.0	90	34.9	[17]
spray	4.4	85	44.7	[18]
sputtering	12.5	95	47.9	[19]
evaporation	25.0	98	32.6	[20]
spray	12.0	93.7	43.5	[1] and this work

Table 3. Comparison of the values of ϕ_{TC} for the ITO films

11. Solar cells fabrication

Solar cells were fabricated using (100) n-type (phosphorous doped) single-crystal silicon wafers of 10 Ω-cm resistivity. Both sides of the wafer were polished. Standard wafer cleaning procedure was used. In order to form the barrier, an 80 nm ITO film with a sheet resistance of 30 $\Omega/\square$ was deposited by spray pyrolysis on the silicon substrate treated in H_2O_2 or SC-1 heated solutions during 10 minutes. The ITO thickness was chosen for an effective antireflection action of the ITO film as shown in Figure 11.

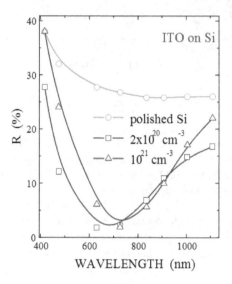

Figure 11. Antireflection action of the 80 nm ITO films for different carrier concentrations.

Metal to form an ohmic contact in the back side of the wafer was deposited on the n^+-layer previously created by diffusion. The device area for measurements was 1-4 cm². Approximately, a 1 μm Cr/Cu/Cr film was evaporated through a metal mask to make a grid pattern of approximately 10 grid lines/cm. After the fabrication, the capacitance-voltage characterization was conducted in order to control the value of the potential barrier. Then the following parameters were measured under AMO and AM1.5 illumination conditions using the solar simulator Spectrolab X25: open circuit voltage V_{oc}, short circuit current I_{sc}, fill factor FF, and efficiency. No attempt was made to optimize the cells for efficiency through a better collection grid. The series resistance (R_s) of the cell was measured using the relationship $R_s=(V-V_{oc})/I_{sc}$ [22], where V is the voltage from the dark I-V characteristic at that point where $I=I_{sc}$. More technological details can be found in [23].

12. Properties of the solar cells

The potential barrier height of the ITO/n-Si solar cells determined from the capacitance-voltage (C-V) characteristics is 0.9 eV. This high value of the potential barrier let us consider such structures as pseudo classical diffusion p-n junctions. Thus, it is possible to expect that the diffusion of holes in the silicon bulk is the main carrier transport mechanism instead of the thermo-ionic emission in the Schottky and the metal/tunnel oxide/semiconductor structures. Moreover, C-V measurements of the potential barrier in structures with a created inversion layer deliver an incorrect the potential barrier [24]. The barrier determined with this method is lower than the actual value.

A straightforward measurement of the temperature dependence of the dark current is, in principle, sufficient to identify a bipolar device in which the thermo-ionic current is negligible compared to the minority-carrier diffusion current J_d (in units of current density). Simple Shockley's analysis of the p-n diode with results for the temperature dependence of the silicon parameters (diffusion length, the diffusion coefficient, the minority carrier life-time, and the intrinsic concentration) [25] shows that

$$J_d = J_{0d} \exp[(qV / kT) - 1] \tag{4}$$

and

$$J_{0d} \propto T^{\gamma} \exp(-E_{g0} / kT) \tag{5}$$

where $\gamma = 2.4$ and $E_{g0} = 1.20$ eV.

From Eq.(5) it can be seen that a plot of $\log(J_{0d}/T^{\gamma})$ versus $1/T$ should yield a straight line, and that the slope of this line should be the energy E_{g0}. In the case of MS and MIS devices this slop must be equal to the value of the barrier φ_b. Usually, the series resistance of the device affects the I-V characteristics at high forward current densities. To prevent this second order effect, we have to measure the J_{sc} vs V_{oc} dependence [22]. The photogenerated current is equal to the saturation photocurrent. For a minority-carrier MIS diode with a thin insulating layer [25]

$$J_{sc} = J_{rg}(V_{oc}) + J_d(V_{oc}) \tag{6}$$

As J_d increases more rapidly with bias than the recombination current density J_{rg}, in the high illumination limit we should have

$$J_{sc} = J_{0d} \exp(qV_{oc} / nkT), \tag{7}$$

where the n factor is very close to 1.

Figure 12 shows the measured dependence of J_{sc} on V_{oc} at room temperature. The value of J_{0d} in (7) was determined by measuring J_{sc} and V_{oc} at different temperatures under illumination with a tungsten lamp. An optical filter was used to prevent the heating of the cell by the infra-red radiation. For each J_{sc} - V_{oc} pair lying in the range where $n \approx 1$, the value $J_{0d}=J_{02}$ was calculated from (7). After correction for the T^{γ} factor appearing in Eq. (5), the J_{0d} values were plotted as a function of reciprocal temperature as shown in the insert of Figure 11.

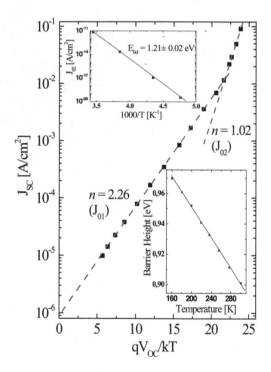

Figure 12. Measured dependence of J_{sc} on V_{oc} at room temperature and calculated dependence of the current density $J_{02}=J_{0d}$ under a high illumination level corrected for the T^{γ} factor as function of reciprocal temperature for ITO/n-Si solar cells. The dependence of the barrier height on temperature is also shown in the insert.

The slope of the J_{02} vs $1/T$ line was found to correspond to energy E_{g0} from Eq. (5). It can be concluded that, for high current densities, the current in the cell is carried almost exclusively by holes injected from the ITO contact and diffusing into the base of the cell. Below we give other independent evidence of the diffusion mechanism.

Output characteristics of the ITO/n-Si solar cell, measured under AM0 and AM1.5 illumination conditions, as well as the calculated dependence of the output power of the cell versus the photocurrent, are shown in Figure 13.

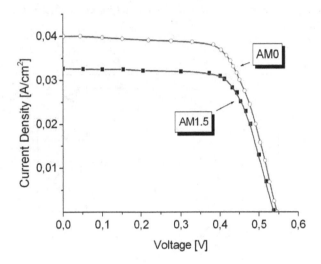

Figure 13. Loading *I-V* characteristics of the ITO/n-Si solar cell measured under AM0 and AM1.5 illumination conditions.

The fill factor (FF) and the efficiency calculated from these characteristics are 0.68 and 10.8% for AM0 illumination conditions, whereas they are 0.68 and 12.1% under the AM1.5 conditions.

We can observe that the parameters of the solar cells fabricated with the silicon wafers treated in hot SC-1 solutions, with an error of ±10%, coincide with those obtained using wafers treated in hot H_2O_2. At the same time, the parameters of the cells fabricated on wafers without these treatments, when the ITO film was deposited on the silicon wafer after a treatment in an HF solution, were significantly lower.

13. Direct evidence of minority carrier injection in ITO-Si solar cells: Bipolar transistor

Since the barrier height exceeds one half of the silicon energy band-gap, the formation of an inversion p-layer at the silicon surface is obvious. To avoid any speculations in this issue and to present the direct evidence of the existence of a minority (hole) carrier transport in ITO-nSi structures, a bipolar transistor structure was fabricated on a 10 Ω-cm monocrystalline silicon substrate, in which the emitter and the collector areas were fabricated using ITO/n-Si junctions, and the ITO film was deposited by the spray technique described above followed by a photolithographic formation of the emitter and the collector areas. The silicon substrate followed the treatment in SC-1 or H_2O_2 solutions described above. An ohmic n⁺-contact (a base) was formed by local diffusion of phosphorous in the silicon substrate. Figure 14 shows the dependence of the collector current versus the collector-base voltage using the emitter current as a parameter as well as the emitter injection efficiency of the ITO/n-Si/ITO transistor [26].

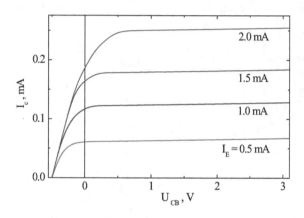

Figure 14. Dependence of the collector current versus the collector-base voltage (the emitter current is used as a parameter). The emitter injection efficiency of the ITO/n-Si/ITO transistor fabricated on a 10 Ω-cm silicon substrate is also shown.

Thus, we obtained an efficiency of 0.2-0.3 even for a non-optimized long base transistor. This makes an obvious evidence of the existence of an inversion layer formed in the ITO/n-Si structures with a barrier height of 0.9 eV. We can also compare our results with the first bipolar transistor based on germanium, in which the existence of an inversion layer on the germanium surface determined a high injection level of minority carriers. In metal-semiconductor contacts, operated as majority carriers' devices, and described by the Schottky theory, the injection ratio does not exceed the value of 10^{-4}. Thus, based on such unipolarity, the fabrication of the bipolar transistor is impossible.

14. Radiation emission from ITO-nSi structures

Figure 15 shows the radiation emission obtained from the ITO/n-Si structures under forward bias [26].

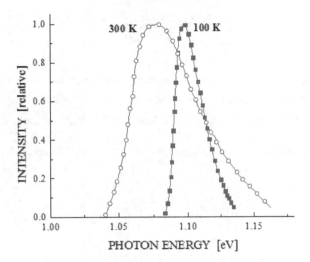

Figure 15. Normalized electroluminescence spectra obtained from an ITO/n-Si structure.

The pumping current density is 500 A/cm².

An intense luminescence was observed from the ITO/n-Si structures. Estimations give an internal quantum efficiency of about 10^{-4}. This is only possible for the case of a high value of the injection coefficient of minority carriers.

15. Minority-carrier injection ratio

The minority-carrier injection ratio or injection coefficient γ is defined as

$$\gamma = \frac{J_p}{J_p + J_n}$$

where J_p is the hole injection current and J_n is the majority-carrier contribution to the forward-biased current in a Schottky barrier.

According to [27]:

$$J_p = \frac{qD_p p_{n0}}{L_p}\left[\exp\left(\frac{qV}{kT}\right) - 1\right], \tag{8}$$

where p_{n0} is the equilibrium hole concentration in the neutral n-region, D_p is the diffusion coefficient, L_p is the diffusion length for holes, and V is the applied forward voltage.

Taking into account that $p_{n0} = n_i^2 / N_d$ where n_i is the intrinsic concentration given by

$$n_i = \left(N_c N_v\right)^{1/2} \exp\left(-\frac{E_g}{2kT}\right) \qquad (9)$$

with N_c, N_v being the effective density of states in the conduction and valence bands, respectively, and E_g is the energy gap. Taking into account that $L_p = (D_p \tau_p)^{1/2}$, $D_p = kT \mu_p / q$, where τ_p is the lifetime of holes in the n region, μ_p the hole mobility in this region, and introducing the term $\mathrm{csch}\left(\dfrac{W_b}{L_p}\right)$ where W_b is the n-base width, for $V >> kT/q$ Eq. (8) can be rewritten,

$$J_p \cong \frac{q D_p^{1/2} N_c N_v}{\tau_p^{1/2} N_d} \exp\left(\frac{qV - E_g}{kT}\right) \mathrm{csch}\left(\frac{W_b}{L_p}\right). \qquad (10)$$

The majority-carrier current is

$$J_n = A^* T^2 \exp\left(-\frac{q\phi_b}{kT}\right)\left[\exp\left(\frac{qV}{kT}\right) - 1\right] \qquad (11)$$

where A^* is the modified Richardson constant and φ_b is the barrier height.

From (10) and (11)

$$\gamma = \frac{J_p}{J_p + J_n} = \left[1 + \frac{A^* T^2 L_p N_d}{q D_p N_c N_v \, \mathrm{csch}\left(\dfrac{W_b}{L_p}\right)} \exp\left(\frac{E_g - q\phi_b}{kT}\right)\right]^{-1}. \qquad (12)$$

Using (12) we try to find out the value of the barrier height for the case of our ITO-nSi solar cells ($N_d = 5 \times 10^{14}$ cm^{-3}, $L_p = 0.4$ mm) that fit to the experimentally obtained value; in this case it was 1.03 V (Figure 16).

Thus, from the experimental data and theoretical estimations the barrier height in ITO-nSi structures is very high and due to the strong inversion condition at the surface of n-silicon.

As the next step, we will explain this phenomenological fact correlating our experimental and estimated results.

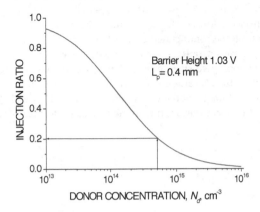

Figure 16. Dependence of the injection ratio on the donor concentration calculated using equation (12) for a barrier height of 1.03 V and a diffusion length of 0.4 mm.

16. Physical model of spray deposited ITO/n-Si solar cells

For this purpose, we have to take into account the following remarks:

1. The thin insulating layer formed at the silicon surface after boiling it in SC-1 or hydrogen-peroxide (H_2O_2) solutions is very thin, about 0.8-1 nm [4-8], and presents negative charge [4]. The thickness of this layer according to [28] can be also 0.68 nm.

2. Below this thickness, there is no limit for the electron flow from the n-type silicon substrate due to the tunnel effect.

3. An additional negative charge can be present at either, the silicon surface or the insulating layer, due to the Al/or Fe contamination introduced during the boiling of the wafers in SC-1 or in the hydrogen-peroxide (H_2O_2) solutions. The origin of this contamination can be the use of non-highly purified H_2O_2 or the dissolving of the Pyrex glass wall of the container by NH_4OH.

4. The band banding due to this inversion condition is formed at the silicon surface *after* the chemical treatment using either, SC-1 or hydrogen-peroxide (H_2O_2) solutions, before the ITO film deposition by spray pyrolysis. Based on published data [9,10], the surface potential (diffusion potential) in the *n*-Si 10 Ω-cm due to acceptor-like surface states can be as high as 0.75 eV how is shown in Figure 2, if the negative charge 8×10^{-8} C/cm^2 is located near the silicon surface.

5. The work function of the ITO films is 4.8 eV [13].

Below, by using Figure 17, we demonstrate the formation of the ITO-nSi structure in two stages. First, after the boiling in SC-1 or hydrogen-peroxide (H_2O_2) solutions, the inversion layer at the silicon surface is formed due to acceptor-like surface states (negative charge due to trapped electrons from the conducting band of the silicon); after the ITO film deposition, and due to the fact that $q\phi_{ITO} < qX_s+(E_g-\Delta)$, some electrons coming from the conduction band of the ITO film, fill non-occupied surface states levels above the semiconductor Fermi level. When the Fermi levels of the ITO and the Si coincide for thermal equilibrium, the value of Δ decreases and the value of the potential barrier $q\phi_b$ increases. This leads to a stronger inversion band bending at the silicon surface due to the increasing surface negative charge.

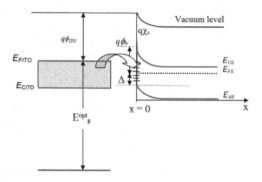

Figure 17. Formation of the inversion layer in the ITO-nSi structure when the ITO film is deposited on the silicon surface with surface barrier $q\phi_b$.

Second, consider the energy band diagram of the ITO-nSi solar cells with an inversion surface layer is shown in Figure 18. Our estimation based on the injection ratio gives a barrier height of $q\phi_b$=1.03 eV. Thus a p^+-inversion layer is formed at the silicon surface.

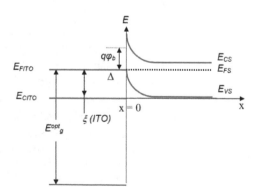

Figure 18. Energy band diagram of the ITO/n-Si solar cells with an inverted surface.

The role of the ITO film is to create an ohmic contact with the $p+$-inversion layer. Furthermore, this film serves as a supplier of holes for the inversion layer. Because of the short distance (~0.1 eV) existing between the top of the silicon valence band (E_{vs}) and the Fermi level (E_{FITO}) of the ITO film, the number of empty states below the Fermi level in ITO is about 10^{19} cm^{-3} [29].

17. Waste-free solar modules

Usually solar cells have a rectangular form and are obtained by cutting them from a circular silicon wafer, given place to a waste of about 41%. We propose a different geometrical approach allowing for the fabrication of circular modules without the waste of silicon. First, the silicon cell fabricated using round wafer is cut in sectors as shown in Figure 19.

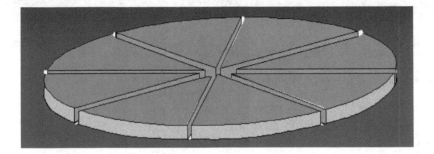

Figure 19. A round solar cell cut in sectors for circular packaging.

Then, the cut sectors are mounted on a plastic base and connected in series. The base has two output electrical contacts. After that, a transparent relief plastic cover is hermetically connected with the base of the module. Solar modules with a rectangular shape can be assembled from sliced solar cells as shown in Figure 20.

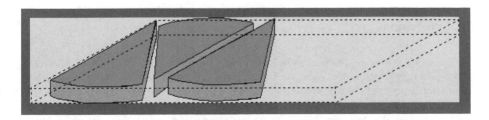

Figure 20. Arrangement of sliced solar cells for rectangular packaging.

Example of waste-free solar modules fabricated on 3 inch silicon wafers is shown in Figure 21.

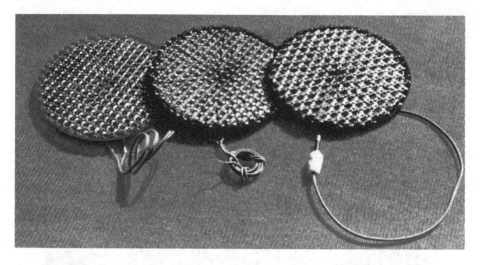

Figure 21. ITO-nSi portable sector-shaped solar modules fabricated by waste-free technology on 3 inch silicon wafers. Each module is assembled from 20 solar cells connected in serial.

The parameters of these modules (under AM1.5 irradiation) are:

Open circuit voltage= 10.8 V;

Short circuit current= 72 mA;

Fill factor= 0.7:

Efficiency=12%.

Using silicon wafers of different diameters, modules with different output parameters can be fabricated. For instance, the module fabricated from a 20 cm diameter silicon wafer cut in 20 sector-shaped solar cells will have an output of 0.5 A and 10.8 V, and a power of 3.8 Watt. Such hermetic modules can be easily assembled in a solar panel without waste of silicon.

18. Conclusions

We discussed some new physical aspects presented by spray deposited ITO-nSi solar cells which are tightly connected with the fabrication technology. It was shown that certain chemical treatment of the silicon wafers in alkaline-peroxide solutions gives rise to the formation of a very thin (0.6-1 nm) insulating layer on the silicon surface. Moreover, a high negative charge, due to acceptor-like surface levels after such chemical treatment, creates a surface $p+$ inversion layer, which leads to a high potential barrier at the silicon surface. After the deposition of the ITO film on the silicon surface, this barrier increases up to 1 eV due to the filling of empty surface states above the silicon Fermi level by electrons coming from the

ITO film. The estimated barrier height agrees very well with the experimentally found value of the minority-carrier injection ratio. A model of the ITO/n-Si solar cells based on the induced surface inversion layer originated by chemical treatments, explains perfectly the transistor effect observed in ITO-nSi-ITO structures, as well as the high level radiation emission from ITO/n-Si devices.

Solar cells based on ITO/n-Si structures are promising for solar energy conversion due to their relativity high output parameters and a low cost fabrication process. Such technological processes are cheap because the lack of high-temperature diffusion processes. Cells with low output parameters after etching of the ITO layer can be used again for an additional solar cells fabrication. We also showed a new waste-free design of solar modules, circular and rectangular, with 40% economy of silicon. Such approach can be successfully applied to any type of solar cells.

Author details

Oleksandr Malik* and F. Javier De la Hidalga-W

*Address all correspondence to: amalik@inaoep.mx

Instituto Nacional de Astrofísica, Óptica y Electrónica (INAOE), Puebla, Mexico

References

[1] Malik O., De la Hidalga-W F.J., Efficient Silicon Solar Cells Fabricated with a Low Cost Spray Technique. In: Rugescu R. (ed.) Solar Energy. Rijeca: InTech; 2010. p. 81-104.

[2] Shewchun J, Buró D, Spitzer M. MIS and SIS Solar Cells. IEEE Trans. Electron Devices, 1980; ED-27(4) 705-715.

[3] Daw A, Datta A, Ash M. On the Open-Circuit Voltage of a Schottky-Barrier MIS Solar Cell. Solid-State Electronics, 1982; 25(12) 1205-1206.

[4] Angermann H. Passivation of Structured P-Type Silicon Interfaces: Effect of Surface Morphology and Wet-Chemical Pre-Treatment. Applied Surface Science, 2008; 254 8067-8074.

[5] Bertagna V et al. Electrochemical Impedance Spectroscopy as a Probe for Wet Chemical Silicon Oxide Characterization. J. Solid State Electrochem, 2001; 5 306-312.

[6] Bertagna V et al. Electrochemistry, a Powerful Tool for the Investigation of the Nanoscale Processes at Silicon Surface. J. of New Materials for Electrochemical Systems. 2006; 9 277-282.

[7]   Neuwald U. Chemical Oxidation of Hydrogen Passivated Si (111) Surfaces in H_2O_2. J. Appl. Phys., 1995; 78(6) 4131-4136.

[8]   Verhaverbeke S, Parker J, McConnell. The Role of HO_2^- in SC-1 Cleaning Solutions. In: Material Research Symposium Proceeding. 1997; 477 47-56.

[9]   Munaka C, Shimizu H. Aluminium-Induced AC Surface Photovoltages in N-Type Silicon Wafers. Semicond. Sci. Technol., 1990; 5 991-993.

[10] Shimizu H, Shin R, Ikeda M. Quantitative Estimation of the Metal-Induced Negative Charge Density in N-Type Silicon Wafers From Measurements of Frequency-Depended AC Surface Photovoltage. Jpn. J. Appl. Phys., 2006; 45 1471-1476.

[11] Pan C, Ma T. Work Function of In_2O_3 Film as Determined From Internal Photoemission. Appl. Phys. Letters, 1980; 37 714-716.

[12] Ginley D., editor. Handbook of Transparent Conductors. New York: Springer; 2010.

[13] Nakasa A. et al. Increase in the Conductivity and Work Function of Pyrosol Indium Tin Oxide by Infrared Irradiation. Thin Solid Films, 2005; 484(1-2) 272-277.

[14] Sato Y, et al. Carrier Density Dependence of Optical Band Gap and Work Function in Sn-Doped In_2O_3 Films, Applied Physics Express, 2010; 3 061101-1/3.

[15] Gouskov L, et al. Sprayed indium tin oxide layers: Optical parameters in the near-IR and evaluation of performance as a transparent antireflecting and conducting coating on GaSb or $Ga_{1-x}Al_xSb$ for IR photodetection. Thin Solid Films, 1983; 99 (4) 365-369.

[16] Vasu V, Subrahmanyam, A. Reaction kinetics of the formation of indium tin oxide films grown by spray pyrolysis. Thin Solid Films, 1990; 193-194 (2) 696-703.

[17] Manifacier J, Fillard J, Bind J. Deposition of In_2O_3-SnO_2 layers on glass substrates using a spraying method. Thin Solid Films, 1981; 77(1-3) 67-80.

[18] Saxena A, et al. Thickness dependence of the electrical and structural properties of In_2O_3:Sn films, Thin Solid Films, 1984; 117(2) 95-100.

[19] Theuwissen A, Declerck G. Optical and electrical properties of reactively d. c. magnetron-sputtered In_2O_3: Sn films. Thin Solid Films, 1984; 121(2) 109-119.

[20] Nath P, Bunshah R. Preparation of In_2O_3 and tin-doped In_2O_3 films by novel activated reactive evaporation technique. Thin Solid Films, 1980; 69(1) 63-68.

[21] Haacke J. New figure of merit for transparent conductors. J. Appl. Phys., 1976; 47 4086- 4089.

[22] Rajkanan K, Shewchun J. A better approach to the evaluation of the series resistance of solar cells. Sol. St. Electron., 1979; 22(2) 193-197.

[23] Malik O, De la Hidalga-W, F.J., Zúñiga-I C, Ruiz-T G. Efficient ITO-Si solar cells and power modules fabricated with a low temperature technology: results and perspectives. J. Non-Cryst. Sol., 2008; 354 2472-2477.

[24] Rhoderick E.H. Metal-semiconductor contacts. Oxford: Clarendon Press; 1978.

[25] Tarr N, Pulfrey D. New experimental evidence for minority-carrier MIS diodes. Appl. Phys. Lett., 1979; 34(4) 295-297.

[26] Malik O, Grimalsky V, Torres-J A, De la Hidalga-W F.J., Room Temperature Electroluminescence from Metal Oxide-Silicon. In: Proceedings of the 16th International Conference on Microelectronics, ICM 2004, December 06-08, Tunisia, Tunis, 471-474.

[27] Buchanan D, Card H. On the Dark Currents in Germanium Schottky-Barrier Photodetectors. IEEE Trans. On Electron Devices, 1982; ED-29(1) 154-157.

[28] Petitdidier S. et al. Growth mechanism and characterization of chemical oxide films produced in peroxide mixtures on Si (100) surfaces. Thin Solid Films, 2005; 476 51–58.

[29] Malik O, Grimalsky V, De la Hidalga-W F.J. Spray deposited heavy doped indium oxide films as an efficient hole supplier in silicon light-emitting diodes. J. Non-Cryst. Sol., 2006; 352 1461-1465.

Chapter 7

Photovoltaic Water Pumping System in Niger

Madougou Saïdou, Kaka Mohamadou and
Sissoko Gregoire

Additional information is available at the end of the chapter

1. Introduction

Solar energy such as photovoltaic is the most promising energy of the non-conventional energy sources which is capable to satisfy the energy needs of the isolated rural areas. It fits perfectly to the decentralization of power generation for the small communities widely dispersed as evidenced the solar pumps whose operation is nowadays proved flawless (World Resources Institute, 1992).

The use of photovoltaic energy to pump water is particularly well suited in the Sahel. This source of energy is free and abundant, but also provides autonomy for many isolated villages of rural areas. The water pumped is stored in the thanks until its use (in the night or during the cloudy days). Locally, there are many companies that manufacture these tanks. These towers do not require special maintenance and are easy to be repair. This chapter presents a study of photovoltaic water pumping in Niger rural areas. We present first the benefits of photovoltaic water pumping; we describe in the second the photovoltaic water pumping system, before present the solar radiation at Niger. Then the sizing of a photovoltaic water pumping system is described and finally, a case study of photovoltaic water pumping is presented before concluding.

2. Benefits of photovoltaic water pumping system

In African countries, many governments are still struggling to meet the basic needs of the population due to lack of availability of electric power system. In rural areas, these needs are summarized in the drinking water, electrification of health centers and irrigation. For lack of resources, these people are resolved to supply water from wells craft of 10 to 100 m and less than 2 m of diameter. Water is the source of life. Its rarity is one of the tragedies of the Sahel. In this area, people have regularly serious problems of drinking water. The water which involves in agricultural and domestic consumption requires dewatering technologies adapted to local conditions. Yet, the solar energy potential is very abundant. The geographical situation of Niger fosters the development and growth of the use of solar energy. Indeed, given the importance of the solar radiation and the duration of sunshine, which excesses eight hours (8 hours) per day throughout the year, some electricity needs can be met by solar energy. This energy source may be beneficial to the most remote areas, especially in water pumping. The use of solar energy can make a meaningful and lasting solution to the drainage in rural areas. The use of photovoltaic solar energy for water pumping is well suited to these regions due to the existence of a potential groundwater quite important especially in desert areas, and a large solar energy potential. Also, this solar energy has the advantage of being clean compared to conventional energy sources which present constraints of remoteness from the mains, fuel transport and periodic maintenance for diesel engines (Jimmy et al 1998, Barlow, 1993). Photovoltaic solar energy may well contribute to meet the energy needs of these populations. Photovoltaic generators are very suitable to relieve the lack of availability of grid. Indeed, with an installation of 1 kW maximum power, we can pump water to heights of about 80 m, or meet irrigation needs of a village (Commission of The European Community, 1985). In addition, these solar generators can be used concomitantly using batteries to power television sets, or for lighting when the pumping system is not loaded (Billerey, 1984).

3. Photovoltaic pumping system

3.1. Description

Photovoltaic pumping system is a standard pump equipped with an electric motor, provided in electrical energy by photovoltaic panels installed on the site (Handbook on Solar Water Pumping, 1984; Fraenkel and al., 1986). This pump is intended to pump water from the basement to make it accessible to users (Fig. 1).

Nowadays, two types of photovoltaic water pumping systems are used: the photovoltaic water pumping with batteries and without batteries. In Niger, it is often used photovoltaic water pumping without batteries, commonly known as "pumping over the sun". Pumping over the sun is simpler and less expensive than with battery system. Instead of batteries, they use a tank to store water until it is used (Fig.2). Hydraulic storage allows overcoming electrical energy storage thus avoiding the use of batteries which have a limited life (6 years compared with 20 to 30 years of photovoltaic panels) and are polluting. However, the method without batteries has some drawbacks and its main fault is to have a flow of water which depends on the average time of the sunlight (A. Hadj, 1999).

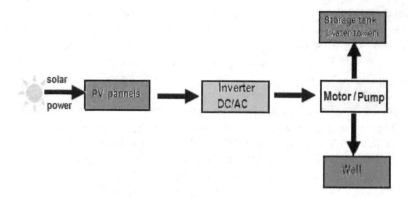

Figure 1. Principe of photovoltaic pumping system.

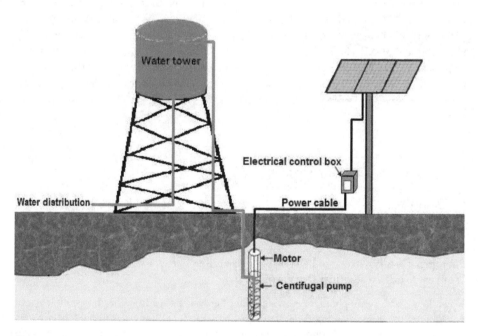

Figure 2. Photovoltaic water pumping without batteries with a tank to store water.

There are two types of pumps to draw water surface: Positive Displacement *Pumps* (volumetric pumps) and centrifugal pumps. Besides to the type of pump, there are two other characteristics at the pumps according to the physical location of the pump in relation to the pumped water: the suction system and stuffer one. They discharge pumps are submerged in water. Their motor is immersed in water with the pump and the discharge pipe placed after the pump can lift water to tens of meters to the storage tank depending to the engine's power. Afterward, the system is connected to a distribution network that delivers water to users.

3.2. Photovoltaic system

To generate the necessary energy to the motor of the pump, solar photovoltaic panels are placed for converting solar energy into electrical energy (Fig.3). As the panels generate a direct current (DC), it is often used DC/AC converter to convert the direct current produced by the solar panels into alternative current (AC) if the motor of the pump is AC. On the other side, if the motor is DC, the device does not need a DC/AC converter. The energy produced by the panels can be used directly or stored. In the case of an application for water pumping, it is more interesting to use the energy to raise the water in a castle that serves as hydraulic energy storage.

To prevent a dysfunction of the pump when it is live on photovoltaic, due to under sizing or over sizing the PV generator, an inverter is used to ensure the proper operation of the PV/ pump system.

Figure 3. Students determining the current-voltage characteristic I = f(V) of solar cell "over the sun" at the University Abdou Moumouni of Niamey (Niger).

4. Solar radiation at Niger

To design a photovoltaic water pumping system, we will need to quantify the available solar energy. Therefore, it is very important to know the solar radiation of the locality. Solar radiation (kW/m^2) is the energy from the sun that reaches the earth. The earth receives a nearly constant of solar radiation at its outer atmosphere. The intensity of solar radiation varies with geographic location (Fig.4).

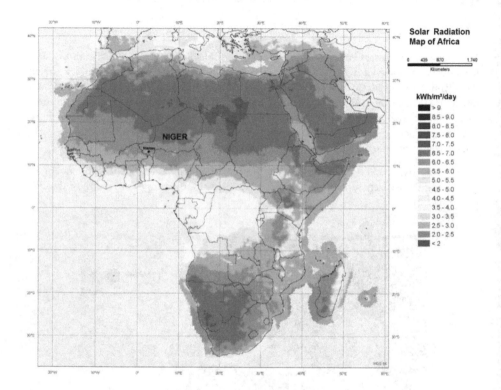

Figure 4. Africa solar radiation map (Source: UNEP, NREL and the Global Environment Facility).

It also varies with the season and time of the day. In Sahel, the solar radiation at the earth's surface is very important (Fig.5).

The most productive hours of sunlight are from 9:00 a.m. to 5:00 p.m. Table 1 gives the daily average time (Hours/day) of the sunlight at Niger (Keita town: Latitude 14.75°N and Longitude 5.76°E).

Month	Jan.	Feb.	Mar.	Ap.	Ma	June	July	**Aug.**	Sept.	Oct.	Nov.	Dec.
Average Time (Hours/day)	9.0	9.3	8.7	8.4	8.8	8.6	8.2	**7.5**	8.1	9.2	9.5	8.9

Table 1. Monthly average time (Hours/days) of the sunlight at Niger (Keita town: Latitude: 14.75°N and Longitude: 5.76°E)

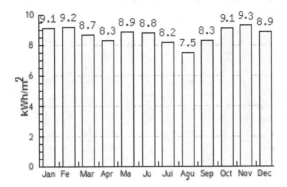

Figure 5. Daily average radiation of sunlight per month at Niger (kWh/m².day).

5. Sizing a photovoltaic water pumping system

Photovoltaic water pump sizing is the determination of the power of the solar generator that will provide the desired amount of water (Alonzo, 2003).

The photovoltaic water pump sizing consists of:

- Assessment of daily water needs of the population to know the rate flow required;
- calculation of hydropower helpful;
- determining of the available solar energy;
- determining of the inclination of the photovoltaic generator which can be placed;
- determination of the month sizing (the month in which the ratio between solar radiation and hydropower is minimum);
- sizing of the PV generator (determination of the required electrical energy);

5.1. Assessment of daily water needs of the population

Determining the water needs of the consumption of a population depends mainly of its lifestyle, the environment and climatic conditions of each region. Drinking, cooking, washing and bathing are the main uses of water for human needs. Animals also need water for their survival. The water use is also essential in the field of agriculture.

Depending on the nature of the users (humans, animals) or the use, the amount of water required for each user or usage are:

- **Population (Humans)**
   - 5 liters /day, for survival;
   - 10 liters /day, for the minimum acceptable;

– 30 liters /day, the normal living conditions in Africa;

• **Animals**

– Cattle	40 liters /day;
– Sheep, Goat	4 liters /day;
– Horse	40 liters /day
– Ass	20 liters /day
– Camel	20 liters /day (reserve for 8 days)

• **Irrigation (agriculture)**

– Crops at the village	60 m³ /day / hectare;
– Rice	100 m³ /day / hectare;
– Cereals	45 m³ /day / hectare;
– Sugar cane	65 m³ /day / hectare;
– Cotton	55 m³ /day / hectare.

There are three standards for the calculation of water requirements:

1. The standard for the minimum amount for survival.

2. The current target of funding agencies: 20 liters/day/person that does not include livestock and gardening.

3. The minimum amount necessary for economic development: 50 liters/day/person. It includes:

20 liters/day/person, for personal needs;

20 liters / person / day, 0.5 unit of cattle per person;

10 liters / day / person, 2 m² for vegetable crops.

In applying the standards 2 and 3, we obtain the following water requirements (Table 2) depending on the size of the village (population).

Population of village (persons)	Standard 2: (20 liters/day/person) (m³)	Standard 3: (50 liters/day/person) (m³)
250	5	12.5
500	**10**	**25**
750	15	37.5
1000	20	50
1500	20	75
2000	40	100

Table 2. Water requirements depending on the size of the village (population).

5.2. Determination of hydropower helpful

The average daily load i.e. hydropower helpful (kWh/day) required is expressed by:

$$E_H = \frac{g^* \rho_a^* Q_a^* TH}{\eta_p^* 3600} = \frac{C_H^* Q_a^* TH}{\eta_p} \qquad (1)$$

Where, g is acceleration of gravity (9.81 m.s^{-2});

ρ_a is water density (1000 kg/m^3);

Q_a is daily water needs (m^3/day);

TH is the total head (m);

η_p is pump system efficiency

The tank capacity is determined by the daily water needs and the autonomy of the system.

5.3. Determining of the available solar energy

The method used is based on the determination of daily mean values of solar radiation available and hydropower necessary.

5.4. Determining of the inclination of the photovoltaic generator which can be placed

The inclination β to the horizontal plane of the photovoltaic panels (PV) must be to maximize the relationship between solar radiation and hydropower necessary. We have chosen 15°N, the latitude of the locality (Keita town: Latitude 14.75°N).

5.5. Determination of the month sizing

The sizing month will be the worst month, i.e., the month that the ratio between solar radiance and hydraulic energy required is minimal. In our case it is the month of August is the month of sizing. In august the average time is 6.5 hours/day.

5.6. Sizing of the PV generator

As the system works all year round, the field is tilted at an angle equal to the latitude of 15 ° N. It was in August that the average number of hours of sunshine is the lowest maximum: 6.5 hours/day. Assuming a 25% loss due to the temperature and dust, the required electrical energy is given by:

$$W_{PV} = \frac{E_H}{Radiance^*(1 - loss)} \qquad (2)$$

6. Case study

A study on the photovoltaic water pumping system in a village at 30 km of Keita (Niger) to meet the water needs of the five hundred (500) persons gave the following summarized results.

Figure 6. Photovoltaic water pumping station with a volumetric pump at Niger.

a. Assessment of daily the water needs: Using the Standard 3: (50 liters/day/person), the water needs rises to 25 m³/day.

b. The average daily load i.e. hydropower helpful (kWh/day) required is given by this expression :

$$E_H = \frac{g^* \rho_a^* Q_a^* TH}{\eta_P^* 3\,600} = \frac{C_H \cdot Q_a^* TH}{\eta_P}$$

With g= 9.81 m.s⁻²

ρ_a= 1000 kg/m³

Q_a= 25 m³/day

TH = 52 m

η_P= 50 %

It provides: E_H =7 085 Wh

c. The available solar energy:

d. Daily average radiation of sunlight varies from 7.5 to 9.3 kWh/m²/day.

To make sure to do a good sizing, we choose the minimum value of average radiance: 7.5 kWh/m2/day.

The inclination to the horizontal plane of the photovoltaic panels is: $\beta = 15°N$.

e.   The sizing month is: August, 6.5 hours/day.

f.   Sizing of the PV generator

Assuming a 25% loss due to the temperature and dust, the required electrical energy is given by this expression:

$$W_{PV} = \frac{E_H}{Radiance^*(1 - loss)}$$

$W_{PV} = 1260$ Wc

The operating point of our photovoltaic field is set at 120 volts due to the characteristics of the inverter. The photovoltaic field will be composed of 10 multiple modules in series. Generator power is 1260 Wc, the rate current is 10.50 A. With photovoltaic panels which have 3.5 A, we will have 3 modules in parallel.

The table 3 shows the summary of case study.

Month of sizing	Radiance (kWh/m²)	Time of sunlight (H)	Loss (%)	E_H(kWh)	W_{PV}(Wc)	Voltage (V)	Current (A)	Configuration of panels
August	7.5	7.5	20	7 085	1260	120	3.5	10 X 3

Table 3. Summary of case study

7. Conclusion

The use of solar energy in Niger, particularly photovoltaic energy, for water pumping is well suited in this arid and semi-arid area due to the existence in this region of an underground water potential, and a large solar energy potential more than 6 kWh/m².

Photovoltaic generators are coupled directly to the pump with a DC/AC converter. Storing water in the tanks avoids additional costs accumulator used to store electrical energy. The case study clearly shows the advantage of photovoltaic pumping system compared to conventional energy one which has many constraints of distance to the power grid, of transportation of fuel, and of periodic maintenance of the engines. The cost of one cubic meter (1 m³) of water pumped by the PV system is more advantageous than others systems. This pumping system constitutes a solution for the water supply of these sparsely populated, remote and isolated areas (Thomas, 1987). With the falling prices of solar panels, this source of energy must be popularized and integrated in the development strategy of these countries.

Author details

Madougou Saïdou[1*], Kaka Mohamadou[1] and Sissoko Gregoire[2]

*Address all correspondence to: nassara01@yahoo.fr

1 Université Abdou Moumouni de Niamey, Niger

2 Université Cheikh Anta Diop de Dakar, Senegal

References

[1] Hadj, A, Arab, F, & Chenlo, K. Mukadam and J.L. Balenzategui, ((1999). Performance of PV Water Pumping Systems". Renewable Energy, N°2, , 18, 191-204.

[2] Barlow, R, & Mcnelis, B. Et Derrick, A, ((1993). Solar Pumping: An Introduction and Update on the Technology, Performance, Costs, and Economics, World Bank Technical Paper Intermediate Technology Publications, London, UK.(168)

[3] Billerey, J. (1984). Le pompage photovoltaïque, GRET, Paris, France.

[4] Commission of The European Community(1985). Renewable Sources of Energy and Village Water Supply in Developing Countries. Directorate General for Development, EEC.

[5] Alonzo, C. (2003). Contribution à l'Optimisation, la Gestion et le Traitement de l'Energie ». Thèse de Doctorat, Toulouse.

[6] Fraenkel, P. (1986). Water Pumping Devices, I.T Power, London, UK.

[7] Handbook on Solar Water Pumping ((1984). Intermediate Technology Power & Partners, eading, 124 p.

[8] Jimmy, R, Thomas, D, Schiller, E, & Sy, B. S. (1998). Le pompage photovoltaïque : manuel de cours à l'intention des ingénieurs et des techniciens. Éditions Multi Mondes. 254 p. 2-89481-006-7

[9] Thomas, M. G. (1987). Water Pumping- The Solar Alternative". Photovoltaic Design, Assistance Center. Sandia National Laboratories. NM 87185, Albuquerque, 58 p.

[10] UNEPNREL and the Global Environment Facility ((2012). www.unep.org/dgef

[11] World Resources Institute(1992). Ressources mondiales 1992-1993 : un guide de l'environnement global, PNUD, Éditions Sciences et Culture, Montréal, Canada.

Permissions

The contributors of this book come from diverse backgrounds, making this book a truly international effort. This book will bring forth new frontiers with its revolutionizing research information and detailed analysis of the nascent developments around the world.

We would like to thank Prof. Dr. Eng. Radu D. Rugescu, for lending his expertise to make the book truly unique. He has played a crucial role in the development of this book. Without his invaluable contribution this book wouldn't have been possible. He has made vital efforts to compile up to date information on the varied aspects of this subject to make this book a valuable addition to the collection of many professionals and students.

This book was conceptualized with the vision of imparting up-to-date information and advanced data in this field. To ensure the same, a matchless editorial board was set up. Every individual on the board went through rigorous rounds of assessment to prove their worth. After which they invested a large part of their time researching and compiling the most relevant data for our readers. Conferences and sessions were held from time to time between the editorial board and the contributing authors to present the data in the most comprehensible form. The editorial team has worked tirelessly to provide valuable and valid information to help people across the globe.

Every chapter published in this book has been scrutinized by our experts. Their significance has been extensively debated. The topics covered herein carry significant findings which will fuel the growth of the discipline. They may even be implemented as practical applications or may be referred to as a beginning point for another development. Chapters in this book were first published by InTech; hereby published with permission under the Creative Commons Attribution License or equivalent.

The editorial board has been involved in producing this book since its inception. They have spent rigorous hours researching and exploring the diverse topics which have resulted in the successful publishing of this book. They have passed on their knowledge of decades through this book. To expedite this challenging task, the publisher supported the team at every step. A small team of assistant editors was also appointed to further simplify the editing procedure and attain best results for the readers.

Our editorial team has been hand-picked from every corner of the world. Their multi-ethnicity adds dynamic inputs to the discussions which result in innovative outcomes. These outcomes are then further discussed with the researchers and contributors who give their valuable feedback and opinion regarding the same. The feedback is then collaborated with the researches and they are edited in a comprehensive manner to aid the understanding of the subject.

Apart from the editorial board, the designing team has also invested a significant amount of their time in understanding the subject and creating the most relevant covers. They scrutinized every image to scout for the most suitable representation of the subject and create an appropriate cover for the book.

The publishing team has been involved in this book since its early stages. They were actively engaged in every process, be it collecting the data, connecting with the contributors or procuring relevant information. The team has been an ardent support to the editorial, designing and production team. Their endless efforts to recruit the best for this project, has resulted in the accomplishment of this book. They are a veteran in the field of academics and their pool of knowledge is as vast as their experience in printing. Their expertise and guidance has proved useful at every step. Their uncompromising quality standards have made this book an exceptional effort. Their encouragement from time to time has been an inspiration for everyone.

The publisher and the editorial board hope that this book will prove to be a valuable piece of knowledge for researchers, students, practitioners and scholars across the globe.

List of Contributors

Onur Taylan and Halil Berberoglu
Department of Mechanical Engineering, The University of Texas at Austin, Austin, TX, USA

Rafael Almanza and Iván Martínez
Universidad Nacional Autónoma de México / Universidad Autónoma del Estado de México, México

Radu D. Rugescu
University "Politehnica" of Bucharest, Romania

P.P. Horley and J. González Hernández
Centro de Investigación en Materiales Avanzados, Chihuahua - Monterrey, México

L. Licea Jiménez, S.A. Pérez García and J. Álvarez Quintana
Centro de Investigación en Materiales Avanzados, Chihuahua - Monterrey, México
GENES Group of Embedded Nanomaterials for Energy Scavenging, Apodaca, México

Yu.V. Vorobiev and R. Ramírez Bon
Centro de Investigación y Estudios Avanzados Unidad Querétaro, Querétaro, México

V.P. Makhniy
Yuri Fedkovych Chernivtsi National University, Chernivtsi, Ukraine

Valentina A. Salomoni, Carmelo E. Majorana and Domenico Mele
Department of Civil, Environmental and Architectural Engineering, University of Padua, Padua, Italy

Giuseppe M. Giannuzzi
ENEA – Agency for New Technologies, Energy and Environment, Thermodynamic Solar Project, CRE Casaccia, Rome, Italy

Rosa Di Maggio and Fabrizio Girardi
Department of Materials Engineering and Industrial Technologies, University of Trento, Trento, Italy

Marco Lucentini
CIRPS, University of Rome "La Sapienza", Rome, Italy

Oleksandr Malik and F. Javier De la Hidalga-W
Instituto Nacional de Astrofísica, Óptica y Electrónica (INAOE), Puebla, Mexico

Madougou Saïdou and Kaka Mohamadou
Université Abdou Moumouni de Niamey, Niger

Sissoko Gregoire
Université Cheikh Anta Diop de Dakar, Senegal

Printed in the USA
CPSIA information can be obtained
at www.ICGtesting.com
JSHW011359221024
72173JS00003B/349